Springer-Verlag Berlin Heidelberg GmbH

G. Jeanette Thorbecke   Vincent K. Tsiagbe

The Biology
of Germinal Centers
in Lymphoid Tissue

Springer

G. Jeanette Thorbecke and Vincent K. Tsiagbe

Department of Pathology
New York University Medical Center
and Kaplan Comprehensive Cancer Center
New York, New York, U.S.A.

ISBN 978-3-662-13143-5

Library of Congress Cataloging-in-Publication data

The biology of germinal centers in lymphoid tissue / [edited by]
G. Jeanette Thorbecke, V. K. Tsiagbe.
    p. cm.-- (Biotechnology intelligence unit)
    Includes bibliographical references and index.
    ISBN 978-3-662-13143-5   ISBN 978-3-662-13141-1 (eBook)
    DOI 10.1007/978-3-662-13141-1

1. Germinal centers.   2. Lymphoid tissue.   I. Thorbecke, G. Jeanette.

2.   II. Tsiagbe, Vincent Kwaku.   III. Series.
[DNLM: 1. Germinal Center—physiology. 2. Lymphoid Tissue—physiology.
  WH 700 B615   1998]
QR185.8.G4B54   1998
616.07'9—dc21
DNLM/DLC
for Library of Congress
                          98-23745
                          CIP

© Springer-Verlag Berlin Heidelberg 1998
Originally published by Springer-Verlag Berlin Heidelberg New York in 1998

The use of general descriptive names, registered names, trademarks, etc. in this publication does not imply, even in the absence of a specific statement, that such names are exempt from the relevant protective laws and regulations and therefore free for general use.

Product liability: The publisher cannot guarantee the accuracy of any information about dosage and application thereof contained in this book. In every individual case the user must check such information by consulting the relevant literature.

Typesetting: R.G. Landes Company, Georgetown, TX, U.S.A.

SPIN 10681646   31/3111 - 5 4 3 2 1 0 - Printed on acid-free paper

This book on germinal centers contains five chapters by leading investigators in the field, each of which covers a special aspect of germinal center biology in an integrated fashion. Germinal centers are foci of antigen-induced rapidly proliferating B lymphocytes, representing sites in mammalian lymphoid tissue where memory B cells are generated. The V regions of the Ig produced by the B cells undergo somatic hypermutation. As a result, the memory cells produced within germinal centers undergo a selection process, ensuring enhanced affinity of their Ig product for the antigen that induced their proliferation.

The impetus for assembling this book stems from the recent surge of interest in germinal centers exhibited by the immunological community. These typical in vivo structures, which, as discussed in chapter 1, are nearly impossible to mimic in tissue culture, have suddenly caught the attention of many scientists. A thorough background information on what is already known about germinal centers, and about the relationship between germinal centers and the rest of the lymphoid tissue, is essential to the design and interpretation of studies on the integral role of these structures in the function of the immune system. Several of the authors contributing chapters to this book have been working on germinal centers for decades and are quite familiar with the development of the major findings in this field as well as the state of the art with respect to present knowledge in this area.

One of the most intriguing questions in molecular immunology today is that of the mechanism of somatic hypermutation. In chapter 4, a major contributor to this area discusses what is known and includes recent findings that are beginning to provide some insight into this process. The relevance of somatic hypermutation to the process of the generation of memory B cells bearing high affinity antigen receptors, and the role of germinal centers in this process, are also discussed in depth in chapter 1. In addition, attention is given to the role of germinal center independent somatic hypermutation in the generation of the virgin B cell repertoire in some mammals, as well as to the presence of a different mechanism of germinal center dependent generation of diversity, gene conversion, which occurs in certain other mammals.

Various aspects of the generation of germinal centers, including the respective roles of the major cellular components, centroblasts, centrocytes, CD4 T cells, follicular and other dendritic cells, and their interactions, are discussed in chapters 1-3. Such interactions lead not only to the proliferation of memory B cells, but also to the selective avoidance of apoptosis by those cells that have succeeded in producing high affinity receptors for the relevant antigen. The roles of many of the critical interacting molecules in the selection process are beginning to be understood and are discussed in detail. Attention is also given to topics such as the identity of germinal center B cell precursors, the homing properties of such cells as influenced by cytokines and chemokines, and the role of antigen localization. The latest information obtained with the use of genetic mouse models is included.

Physiological and pathological conditions in which germinal centers malfunction, including aging, genetic abnormalities and viral infections, are covered in chapters 1 and 3. Because of the tremendous importance of conditions in which malignant transformation of germinal center components has occurred, chapter 5 is completely devoted to a discussion of lymphomas of germinal center origin. The knowledge gained from studies on germinal center derived lymphomas in mice is included to emphasize how understanding of the biology of germinal centers could be crucial for insight into this common form of lymphoma in man.

The inspiration that sustained us throughout the endeavor of writing this book came from interaction with many colleagues in pleasant surroundings at the now traditional triannual "Germinal Center" meetings. We are grateful for the congenial exchange of information and for the enthusiasm about this subject that has finally prevailed.

G. Jeanette Thorbecke and Vincent K. Tsiagbe
May 1, 1998

CONTENTS

EDITORS

V.K. Tsiagbe, Ph.D.
Department of Pathology
New York University Medical Center *and*
Kaplan Comprehensive Cancer Center
New York, New York, U.S.A.
Chapters 1, 5

G.J. Thorbecke, M.D., Ph.D.
Department of Pathology
New York University Medical Center *and*
Kaplan Comprehensive Cancer Center
New York, New York, U.S.A.
Chapters 1, 5

CONTRIBUTORS

G.F. Burton, Ph.D.
Department of Microbiology
   and Immunology
Medical College of Virginia
Virginia Commonwealth University
Richmond, Virginia, U.S.A.
Chapter 2

G.S. Erianne
Department of Pathology
   and Kaplan Comprehensive
   Cancer Center
New York University
   School of Medicine
New York, New York, U.S.A.
Chapter 5

G. Inghirami, M.D.
Department of Pathology
   and Kaplan Comprehensive
   Cancer Center
New York University
   School of Medicine
New York, New York, U.S.A.
Chapter 5

R.A. Insel, M.D.
Professor of Pediatrics,
   Microbiology and Immunology
University of Rochester
   School of Medicine
Rochester, New York, U.S.A.
Chapter 3

M.H. Nahm, M.D.
Professor of Pediatrics,
   Medicine, Pathology
   and Laboratory Medicine
University of Rochester
   School of Medicine
Rochester, New York, U.S.A.
Chapter 3

N.M. Ponzio, Ph.D.
Department of Pathology
   and Laboratory Medicine
University of Medicine
   and Dentistry
New Jersey Medical School
Newark, New Jersey, U.S.A.
Chapter 5

D. Qin, M.D., Ph.D.
Department of Microbiology
  and Immunology
Medical College of Virginia
Virginia Commonwealth
  University
Richmond, Virginia, U.S.A.
Chapter 2

U. Storb, M.D.
Department of Molecular Genetics
  and Cell Biology
University of Chicago
Chicago, Illinois, U.S.A.
Chapter 4

A.K. Szakal, Ph.D.
Professor, Department of Anatomy
Division of Immunobiology
Medical College of Virginia
Virginia Commonwealth
  University
Richmond, Virginia, U.S.A.
Chapter 2

J.G. Tew, Ph.D.
Department of Microbiology
  and Immunology
Medical College of Virginia
Virginia Commonwealth University
Richmond, Virginia, U.S.A.
Chapter 2

J. Wu, M.D.
Department of Microbiology
  and Immunology
Medical College of Virginia
Virginia Commonwealth University
Richmond, Virginia, U.S.A.
Chapter 2

D.J. Zhang, M.D.
Department of Pathology
  and Kaplan Comprehensive
  Cancer Center
New York University
  School of Medicine
New York, New York, U.S.A.
Chapter 5

ABBREVIATIONS

Ab	antibody
AFC	antibody forming cell
Ag	antigen
AIDS	acquired immunodeficiency syndrome
APC	antigen presenting cell
ARR	antigen retaining reticulum
ATC	antigen transport cell
BL	Burkitt's lymphoma
BLR1	Burkitt's lymphoma receptor 1
BSAP	B cell-specific activator protein
CD	cluster of differentiation
CDR	complementarity determining region
CLL	chronic lymphocytic leukemia
CR	complement receptor
DCL	diffuse large-cell lymphoma
DLCL	diffuse large B cell lymphoma
EBV	Epstein-Barr virus
FCCL	follicular center cell lymphoma
FDC	follicular dendritic cells
FR	framework region
GC	germinal center
GCK	germinal center kinase
HD	Hodgkin's disease
HGF/SF	hepatocyte growth factor/scatter factor
HIV	human immunodeficiency virus
HRP	horse radish peroxidase
Ig	immunoglobulin
LC	large cell
LN	lymph node
MALT	mucosa-associated lymphoid tissue
MC	mixed cell
MCL	mantle cell lymphoma
NLPHD	nodular lymphocyte predominant Hodgkin's disease
PALS	periarteriolar lymphatic sheath
PCR	polymerase chain reaction
PHA	phytohemagglutinin P
PNA	peanut agglutinin
PWM	pokeweed mitogen
RAG	recombination activating gene
RCS	reticulum cell sarcoma
RS	Reed-Sternberg
SAC	*Staphylococcus aureus* type C
SC	small cell
SCID	severe combined immunodeficiency disease
V	variable

Overview of Germinal Center Function and Structure in Normal and Genetically Engineered Mice

V.K. Tsiagbe and G.J. Thorbecke

Definition and Description of Germinal Centers

Induction, Overall Structure and Cellular Compartments of Germinal Centers

Germinal centers (GCs) are round to oval shaped areas, located eccentrically in primary B cell follicles, and filled to a variable extent with rapidly proliferating lymphoid blast cells of B cell origin. They become recognizable in draining lymph nodes within 4-5 days after a local antigen injection and have their peak development by 7-10 days. If the antigen inducing their formation was injected without a strong adjuvant, the GCs begin to dissociate by 15 days and progressively lose their proliferative activity by 21-28 days, leaving behind an area in the primary follicle that is less cellular than the rest of the follicle to indicate the prior activity site. GCs are localized on the medullary side of the primary B cell follicles in lymph nodes, on the submucosal side of follicles in Peyer's patches (away from the epithelium which cover the "domes"), and at bifurcations of the arterioles traversing the white pulp in the spleen. Because of the compression caused by the bulging GC, the remainder of the follicle takes on the appearance of a densely populated B cell cap or "mantle zone." Small recirculating B cells localize in these areas as in primary follicles.[1-3] In some species, such as the chicken, GCs are not

The Biology of Germinal Centers in Lymphoid Tissue, edited by G. Jeanette Thorbecke and Vincent K. Tsiagbe. © 1998 Springer-Verlag and R.G. Landes Company.

accompanied by dense compressed areas of small B cells. Such naked GCs are also found in mammals, particularly under conditions in which there is a lack of recirculating B cells.

In lymph nodes, B cell follicles are at the cortical end of the paracortical cords, as described by Gret et al,[4] which represent the unit structure of the paracortex and the connection between follicle and medullary cords. The "corridors" of these paracortical cords contain sessile interdigitating cells and migrating lymphoid cells such that APC-T cell and T-B cell interactions occur here in a microenvironment surrounded by a network of fibroblastic cells and extracellular matrix. The directions in which T and B cells, emerging from high endothelial postcapillary venules into these corridors, migrate is likely to be determined by the presence of chemotactic factors, possibly attached to extracellular matrix, and the differential expression of receptors for these factors by the lymphoid cells (as suggested by Gretz et al[4]). B cells, which differentiate into antibody secreting cells after their interaction with antigen, and T cells appear to migrate towards the medulla, whereas naive B cells migrate into follicles.[1] Most activated B cells are excluded from entering follicles.[5,6] GCs are therefore produced either from naive B cells that become activated within follicles or from activated B cells which, contrary to their fellow activated B cells, migrate into follicles after activation.

In the spleen, the (antigen + helper T cell) activated B cells migrate through the marginal zone bridging channels from the outer "T cell zone" of the periarteriolar lymphatic sheaths (PALS) into the red pulp, where they remain located as foci of antibody secreting cells around the penicilli arterioles (Fig. 1.1[7]). The peak development of these foci coincides with peak antibody formation in the red pulp on days 4-5 after an intravenous injection of antigen.[7-9] Very few of the activated B cells appear involved in GC formation, which first becomes visible around day 4 after antigen injection. When a previously primed animal is challenged with the same antigen, the antibody secreting immature plasma cell foci develop much faster and contain many more cells.[8,10-12] As shown elegantly by Toellner et al,[12] the antigen specific B cells are primarily found in marginal zones prior to the recall injection of antigen, whereas within 12-24 hours after challenge they are found at the junction between the outer PALS region and the follicle, indicating that these cells had been recruited into that area. The antigen specific B cells exhibit mitotic activity, as also shown by incorporation of BUDR,[12] and are seen to move into the red pulp adjacent to the T zone of the PALS. These findings are

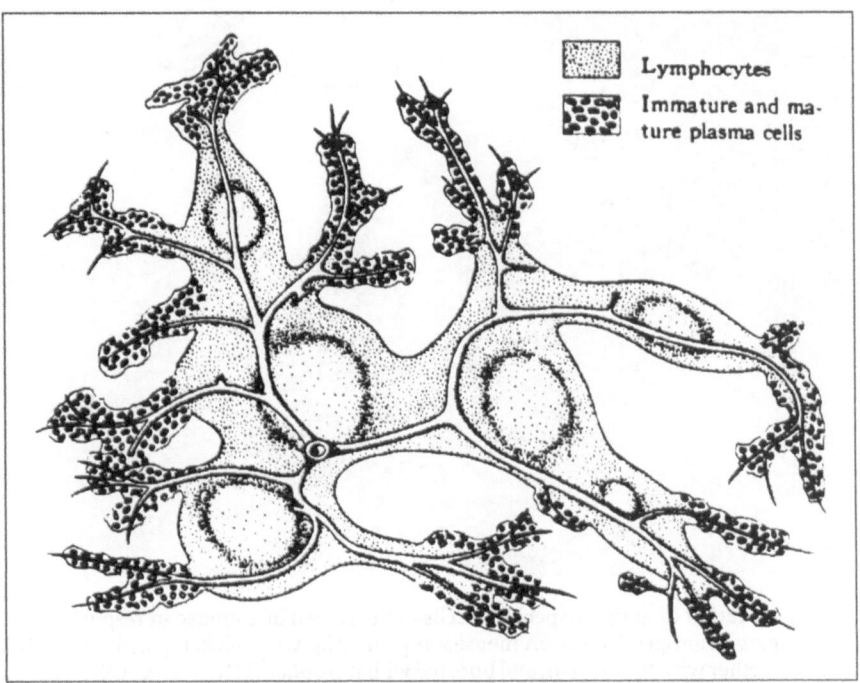

Fig. 1.1. Relationship between white pulp with its GCs and plasma cell aggregates around penicilli arterioles in the red pulp. Reconstruction drawing of a rabbit spleen taken 7 days after the third intravenous injection of *S. paratyphus* vaccine. The graphic reconstruction was made from projections of microscopic views of serial sections stained with methylgreen pyronin. Reprinted with permission from Thorbecke GJ et al, J Infect Dis 1956; 98:157-171.

in total agreement with the older observations that peak antibody formation and immature plasma cell proliferation during the secondary response to an intravenously injected antigen in rabbits is on days 2-3 in the red, rather than the white, pulp of the spleen[8,10] and that these plasma cells do indeed represent the antibody containing cells.[11] The spacial relationship between GCs near the branching place of arterioles in the white pulp (PALS) and the antibody containing cell foci around penicilli arterioles in the red pulp during the secondary response in the spleen is illustrated in Figures 1.1 and 1.2.

Initially, GCs consist of a dense area of rapidly proliferating large blast cells which express low density surface Ig.[13] Liu et al[14] estimate that, according to the number of blast cells present in follicles within a few days after injection of antigen, the cell cycle time in the dark zone must be about 6 hours. During the early stage of proliferation each follicle contains $1-2 \times 10^4$ B cell blasts.[14] These areas arise

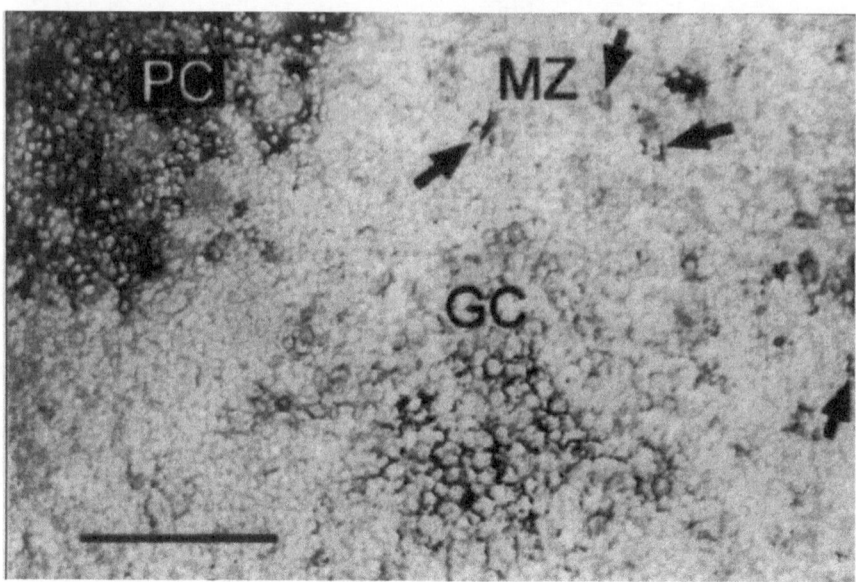

Fig. 1.2. Location of antigen-specific B cells in the spleen of a mouse in response to i.v. challenge with antigen. This B10A mouse was primed i.p. with chicken γ-globulin (CGG), given together with *B. pertussis*, and boosted with nitrophenyl (NP)-CGG. CGG-specific memory B cells are shown in blue (arrow) in the marginal zone (MZ); NP-specific cells are shown in brown. The edge of a red pulp focus of plasma cells (PC) is shown (top left). Immune complexes (black) are present on the FDCs in the GC in the center. Bar = 100 μm. Reprinted with permission from Toellner KM et al, J Exp Med 1996; 183:2303-2312. See color figure, page 235.

from 1-3 precursor cells and represent, therefore, clonal expansions of antigen specific B cells.[14-17] A network of dendritic cells, follicular dendritic cells (FDCs), extends throughout the GCs. At later stages of GC development, particularly in the tonsil where chronic stimulation of GC proliferation occurs, different zones can be detected within the GCs on the basis of blast cell density and FDC staining characteristics, most clearly described and illustrated by MacLennan and coworkers[18,19] (Fig. 1.3). These investigators[20] also made a careful study of the morphology of GCs in human tonsils as compared to lymph nodes. While "light" and "dark" zones reflecting the density of blast cells were detected in both, a division of the light zones into basal, apical and outer zones on the basis of high or low CD23 expression on the FDC network could only be made in the tonsil, as CD23$^+$ FDCs extended into the dark zone as well as the follicular mantle in lymph node GCs. Figure 1.4 (taken from Kosco et al[21]) shows the relationship between the FDC network and centroblasts in murine lymph nodes obtained at different days after immunization. The

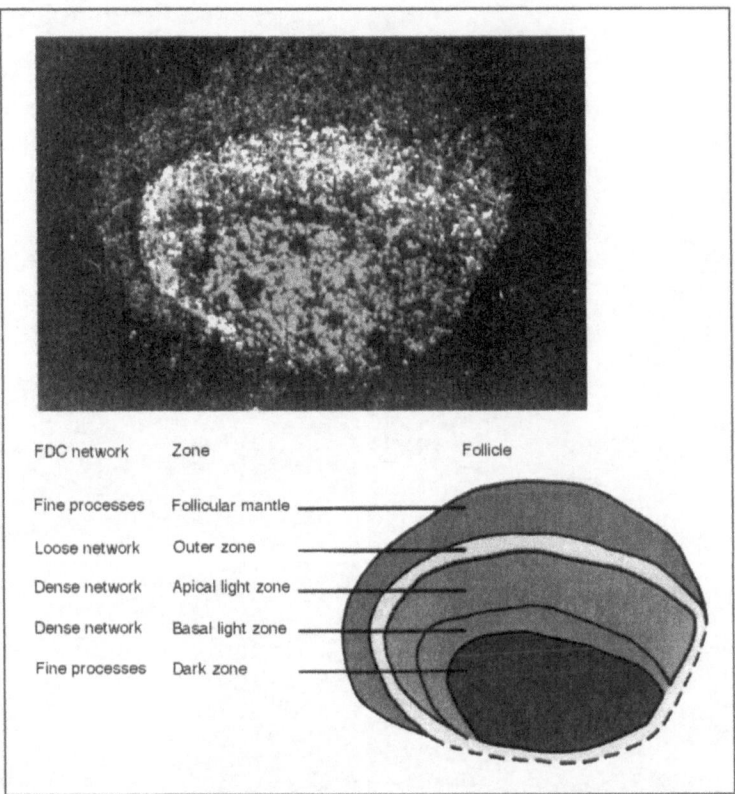

FDC network	Zone	Follicle
Fine processes	Follicular mantle	
Loose network	Outer zone	
Dense network	Apical light zone	
Dense network	Basal light zone	
Fine processes	Dark zone	

Fig. 1.3. Polarization of a germinal center in human tonsil, at a time when B cell blasts (in exponential growth) have filled the follicular dendritic cell (FDC) network. Upper part: A secondary follicle with a well-developed GC is shown in which the zonal pattern, typical of the tonsil GC, is apparent. Green immunofluorescence identifies weak CD23 expression by B cells in the follicular mantle and strong CD23 expression by FDCs in the apical light zone. Red immunofluorescence shows the heavy concentration of cells in cell cycle in the dark zone identified by the Ki67 monoclonal antibody, which identifies dividing cells. The unstained area between the follicular mantle and apical light zone is the outer zone and that between the apical light zone and the dark zone is the basal light zone. Both the outer zone and basal light zone contain occasional Ki67[+] cells. Lower part: These regions are shown with a color key. Reprinted with permission from Liu YJ et al, Immunol Today 1992; 13:18. See color figure, page 236.

dividing cells demarcate the dark zone and the light zone contains a denser FDC network, but basal and apical light zones are not readily distinguished. The dark zone (or base) is furthest away from the follicular mantle zone and closest to the T cell zone in the spleen or paracortex in the lymph node. It contains tightly packed large blast cells, the centroblasts, and has by far the highest mitotic activity.[22]

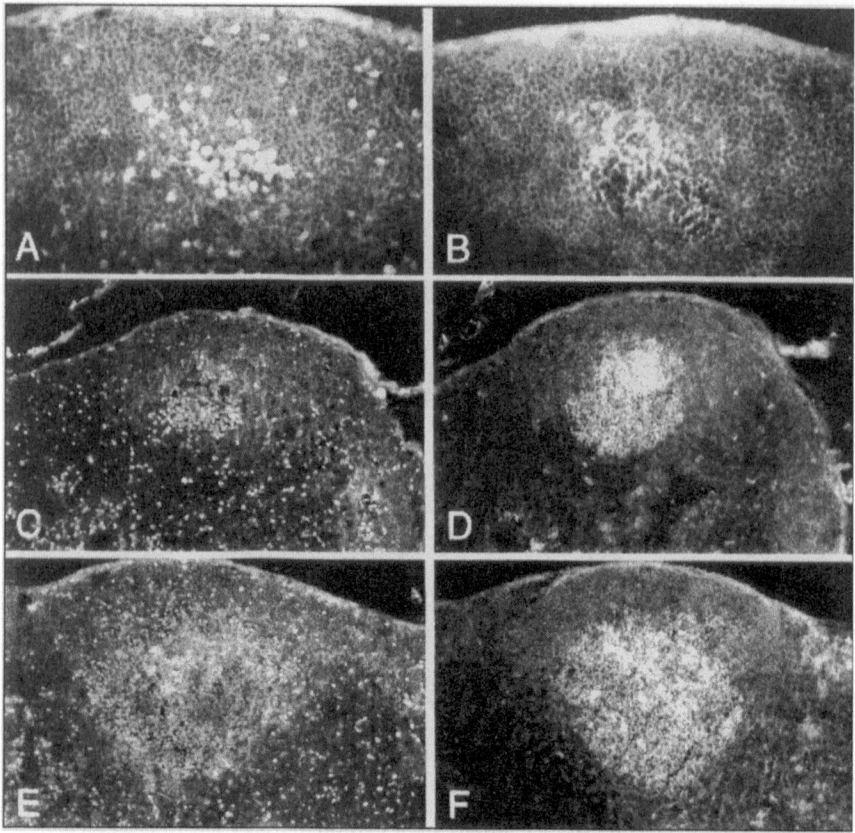

Fig. 1.4. Relationship between dividing cells, PNA$^+$ cells, and the FDC network during the GC reaction. Serial cryosections were prepared from murine lymph nodes obtained 4, 6 and 9 days post primary immunization with aluminum-precipitated ovalbumin. In all panels, IgM$^+$ cells are red. In the left panels (A,C,E), green nuclei correspond to Ki67$^+$ dividing cells. In right panels (B,D,F), green cells correspond to peanut agglutinin (PNA)$^+$ GC cells. The FDC network in the light zone becomes visible on days 6-9, stained red due to immune complexes (ICs) recognized by the anti-IgM reagent. Note that, by day 4, Ki67$^+$ nuclei localize as a small focus of dividing cells in the IgM$^+$ primary follicle. These cells are PNA$^+$ (panel B) and within the FDC network. By day 6, the GC size is expanding and the majority of the Ki67$^+$ nuclei (panel C) are limited to the dark zone, while the FDC network is organized mainly in the light zone (panel D). By day 9, the GC has reached full proportion and many of the cells within the FDC network have Ki67$^+$ nuclei (panels E & F). Magnification: day 4 A-B: x250; day 6 C-D; day 9 E-F: x90. Reprinted with permission from Kosco-Vilbois MH et al, Immunol Today 1997; 18:225-230. Photomicrographs kindly provided by Dr. M. Kosco-Vilbois. See color figure, page 237.

The basal and apical light zones are much less dense and contain smaller blasts or centrocytes, which probably represent the differentiation products of centroblasts. Indeed, DNA-labeling studies suggest that centrocytes are continually replenished from the proliferating pool of centroblasts.[14,22] In human and mouse GCs, both centroblasts and centrocytes have high affinity for peanut agglutinin (PNA;[23] see also Fig. 1.4), while in the rabbit, wheatgerm agglutinin has been used to stain GCs.[24] Such staining with peroxidase conjugated lectins can be extremely useful in the study of size and frequency of GC formation.[25] Some CD4[+] T cells are found in the light zones, particularly in the apical light zone.[26,27] There is a sharp demarcation around the whole follicle center, called the outer zone by Liu et al,[28] which is like a sinusoid containing a mixed cell type, the lining of which is also stained by peanut agglutinin (see Linton et al[29]). Macrophages, containing in their cytoplasm nuclear debris ("tingible bodies") from apoptotic B cells that have recently been in cell cycle,[30] are located throughout GCs, but their number is highest in the basal light zones where the highest rate of apoptosis is seen.[18]

Centroblasts, Centrocytes and Their Relatives

The expression of surface Ig on the rapidly proliferating centroblasts in the dark zone of GCs is low to undetectable. Centrocytes are either μ^+, $\delta^{lo/-}$ or exhibit other Ig isotypes, reflecting the Ig isotype switching which occurs in these cells.[31-33] As shown in Table 1.1, like other B cells, GC cells bear the B cell specific form of CD45 (mouse;[31] rat;[34] man[35]) as well as a variety of other B cell antigens, such as CD19, CD20, CD21 (CR-2), CD22, CD40.[36-38] Like other activated cells, they lack the "homing" receptor mel-14 (L-selectin or leu-8).[39-42] Human GC B cells exhibit several activation antigens, including CALLA (CD10),[37] the transferrin receptor (CD71),[43] Bac-1[44] and carboxypeptidase-M.[45] Adhesion proteins important for cellular interactions are highly expressed on GC B cells, in particular LFA-1 and LFA-3,[38] as well as ICAM-1.[46]

The cell surface marker characteristics compiled in Tables 1.1a and 1.1b represent the staining properties of GC T and B cells in situ (Table 1.1a) and of B cell subpopulations isolated from human tonsil suspensions (Table 1.1b[47,48]). Follicular mantle zone B cells in humans (but not in the mouse!) are CD5[+], while both GC and marginal zone B cells are CD5[-]. In contrast to GC B cells, resting marginal zone B cells in humans are CD77[-], CD10[-], CD38[-], CD71[-], CD62c[+] (L-selectin), IgD[+], CD23[+] and CD44[+].[49] CD38 cannot be used as a marker for

Table 1.1a. Expression of various cell surface markers on GC T and B cells in situ

GC Zone	B Cell Markers							T Cell Markers			
	CD20 and CD40$	10* 77	38*	39	B7-1	B7-2	leu8 44	CD4	40L	28 CTLA4	8
Apical Light	++	+	+	–	–	++	Lo	+	+	+	–
Basal Light	++	+	+	–	+	+	Lo	+	–	+	–
Basal Dark	++	++	++	–	++	–	–	–	–	–	–
Follicular Mantle	++	–	Lo	+	–	–	+	–	–	–	–

Designation: – = < 10%; + = 10-50%; ++ = >50% for CD20[+] cells in the case of B cell markers and for CD4[+] cells in the case of T cell markers. Lo designates dull staining.

* The distribution for CD10 and CD38 only applies to human GC cells.

$ According to a recent study by Tom L. Feldbush (personal communication) on murine GC with double staining for PNA binding and various other markers, murine GC cells (from 1-2 week after immunization) are CD38[-], CD43[-], BLA[-], IgD[-], Fas[+], CD40[lo], CD22[+], CR1/2[+], CR1[lo], CD23[lo], VLA[lo], LFA[+], ICAM-1[+], HSA[hi], and B7-2[lo-med+].

Highest mitotic activity is found in the basal dark zone centroblasts; moderate mitotic activity is seen in the large centrocytes of the basal light zone.

Highest apoptotic activity is found in the large centrocytes of the basal light zone; moderately high apoptosis is also present in the small centrocytes of the apical light zone.

The apical light zone contains the CD23 rich FDCs, whereas the basal light zone contains the CD54 rich FDCs.

Data are from Vyth-Dreese et al;[56] Hardie et al;[18] Feuillard et al.[54]

Table 1.1b. Apoptosis and cellular markers of human tonsil subpopulations

Cell Fraction	Cell Surface Marker					Apoptosis Related Protein				
	CD38	CD39	IgD	CD23	CD77	Bcl-2	Fas	c-Myc	P53	Bax
bm1 foll. mantle	–	+	+	–	–	+	–	–	–	–
bm2 foll. mantle	–	+	+	+	–	+	–	–	–	–
bm3 centroblasts	+	–	–	–	+	–	++	++	+	+
bm4 centrocytes	+	–	–	–	–	–	++	w+	++	++
bm5 memory B	–	+	–	–	–	+	–	w+	–	–

w+ = weakly positive. Data from Liu et al.[47,48]

murine GC B cells, as it is expressed on follicular mantle B and down-regulated on GC B cells in this species.[50] CD10 is also restricted to humans as a GC cell marker, as it is not expressed on murine lymphoid cells.[51] The majority of GC B cells are CD23[-], although a variable content of IgD[+]CD23[+] cells has been reported to occur in the basal dark zone of GCs.[52] CD44 is highly expressed on memory B cells,[53] but absent from centroblasts. Feuillard et al,[54] studying the relative expression of some surface markers on centroblasts and centrocytes, find CD44 absent from centroblasts and weakly expressed on centrocytes. Expression of CD77 is usually higher on centroblasts than on centrocytes. The expression of B7-1 and B7-2 (the ligands for CD28 and CTLA-4) varies between centroblasts and centrocytes is in the mouse; the highest density of B7-1 is on centroblasts and the highest expression of B7-2 on centrocytes in the light zone.[55,56] A few surface antigens have been described that, within normal peripheral lymphoid tissue, are relatively specific for GC B cells, including "BLA",[37,57] CD77 (globotriaosyl ceramide[58,59]), a globoside detected by Ab HJ6,[60] "EAA-B" (a human endothelial activation antigen[61]) and "LN-1".[62] In contrast, an antigen like "4KB51" is present in marginal zone and follicular mantle B cells, but not on GC cells (also positive in all hairy cell lymphomas).[63]

GC B cells combine an absence of Bcl-2,[64] with high expression of Fas(CD95),[65-67] MCL-1,[68] c-Myc,[69] p53[69] and Bax.[70] The inverse of this relationship between the expression of these apoptosis related proteins is seen in mantle zone B lymphocytes. With respect to other intracellular proteins, it is of interest that RAG-1 and RAG-2 are detected in GC B cells.[71,72] Such a reexpression of molecules that are expressed in pro-B and pre-B cells, but absent from mature B cells, has also been observed after stimulation of B cells with LPS or anti-CD40 and IL-4 in vitro[73] and is found in certain B cell lymphomas (see chapter 5). A similar situation exists with respect to expression of telomerase, which is 100- to 1000-fold higher in GC cells than in naive or memory B cells.[74,75] Thus, GC cells are particularly well-equipped to undergo repeated rounds of proliferation without suffering a shortening of their telomeres. Augmented expression in GC centroblasts of 8-oxoguanine DNA glycosylase, which was suggested to be of possible relevance to the process of somatic hypermutation, has recently been detected by subtraction hybridization.[75a]

Bcl-6 is a human 92-98 kDa multiple zinc-finger containing phosphoprotein that is highly expressed in various non-Hodgkin's lymphomas, frequently associated with a breakpoint and/or somatic mutations in 3q27 (see chapter 5). Staining for Bcl-6 has shown it to be a nuclear protein, that, in human lymphoid tissue, is primarily located in GC centroblasts and centrocytes.[76-78] While activation causes a decrease of the Bcl-6 mRNA in resting murine B cells, activated GC cells maintain high mRNA and protein levels of Bcl-6; they contain much higher Bcl-6 protein levels than do other B cells.[80,81] The initial stimuli leading to this high expression in GC cells have not yet been elucidated. Induction of differentiation of the Bcl-6 containing GC cells causes the disappearance of this protein such that plasma cells fail to express it.[79] This downregulation of the Bcl-6 appears to have two components: one is the decrease in mRNA induced by triggering of surface CD40, and the other the destabilization of the Bcl-6 protein through phosphorylation by kinases activated through triggering of surface Ig (R. Della-Favera, personal communication). Other transcription factors that are highly expressed in GC B cells include Spi-B (of the *Ets* family)[82] and A-Myb (also high in Burkitt's lymphoma lines).[83]

Using subtractive cDNA hybridization techniques, Christoph et al[84] have isolated a novel gene (M17) that is expressed in PNA+ (GC) but not in PNA− splenic B cells from immunized mice. M17 shows homology to lipid binding amphipathic helices of the membrane associated molecules C9 and cytolysin. On the basis of this homology, the authors suggest that although M17 codes for a cytoplasmic protein, it may have a lytic role in programmed cell death. With similar techniques, another gene (BL44) has also been found to be preferentially expressed in human GC B cells.[85] BL44 codes for a serine/threonine protein kinase, referred to as GC kinase. Even though GC kinase is expressed in other organs, within lymphoid tissue it is preferentially expressed in GC B cells (and also in the Burkitt's lymphoma line RAMOS). GC kinase specifically activates a mammalian stress-activated protein kinase (SAPK) which is also highly expressed in GC B cells.[86] Both of these kinases are induced by the cytokine TNF-α. The preferential expression of these proteins in GC as compared to other B cells suggests the activation of specific signal transduction pathways in these cells.

Precursors for GC B cells are derived from bone marrow, as is true for most B cells. Bone marrow derivation is confirmed by the reappearance of GCs in lethally irradiated mice reconstituted by bone marrow cells.[87,88] Each GC develops oligoclonally from 1-3 precursor cells,[15] recruited into GCs from the recirculating pool.[89] Several lines of evidence suggest that immature B cells can serve as GC cell precursors. In the chicken, the large immature cells, and not the small high density cells, home to GCs after i.v. injection.[90,91] In the rabbit, large cells from the appendix repopulate the splenic GCs.[92,93] Cell fractionation experiments of VonderHeide and Hunt[94,95] suggest that GC precursors are in the IgD− population, while those of Seijen et al[96] indicate that the GC precursors are in the IgD+ pool. Our own experiments suggest that the GC precursors are normally IgD+, but that the cells do not need to express IgD to produce GCs. When oligomeric IgD is injected before or at the time of antigen injection, IgD receptors (IgD-R) are induced on $CD4^+$ T cells, leading to enhanced memory and GC formation.[97,98] On the other hand, injection of monomeric IgD, at the same time as primary antigen, prevents both the induction of IgD-R on T cells as well as the generation of early B cell memory.[99] Since the effects of IgD injection are not seen in $IgD^{-/-}$ mice, it stands to reason that only $sIgD^+$ B cells are involved in the interaction with $IgD-R^+$ T cells. Nonetheless, like mice treated with anti-IgD from birth, in which IgD^+ cells are lacking, $IgD^{-/-}$ mice are capable of producing GCs and secondary antibody responses.[100-103] The above findings, taken together, suggest that GC B cells are normally derived from antigen-stimulated IgD^+ cells, which become IgD^- after activation either within or outside GCs, but can also be derived from $IgM^+ IgD^{-/-}$ cells, when IgD expression by the B cells is inhibited or genetically absent. The extra impetus provided by sIgD to B cells via $IgD-R^+$ T cells probably facilitates and/or accelerates GC and memory B cell formation, but is not essential.

Not all unprimed (naive) peripheral B cells are equally capable of inducing GCs in severe combined immunodeficient (SCID) hosts. In reconstitution experiments with SCID mouse recipients, immunized after repopulation with excess carrier-primed $CD4^+$ T and purified naive B cell subpopulations, CD24 (HSA, J11D)^{lo} Ia^+ splenic B cells generate many GCs in the spleens of the SCID mice, while $CD24^{hi}$ B cells do not (Fig. 1.5[29]). The $CD24^{lo}$ B cells are mostly IgM^+ (>95%), IgD^{hi}, and express normal levels of CD45 (B220). Similar $CD24^{lo}$ precursors of GC B cells are found in the spleen from athymic mice, suggesting that these cells are not the product of T-dependent cross reactive antigen priming. Although numerous Ig-containing cell foci are found in the recipients of $CD24^{hi}$ splenic or $CD5^+$ peritoneal B cells, neither of these B cell populations gives rise to significant GC formation (Fig. 1.5[29]). However, in cell transfer studies conducted by Kroese et al,[104] peritoneal cells (mostly $CD5^+$) were found to generate few but some definite GCs. In previous studies, Linton et al[105] have shown that, upon antigen stimulation in SCID recipients, $CD24^{lo}$ Ia^+ spleen cells from nonimmune mice give rise to memory B cells and few primary Ab forming cells, while the $CD24^{hi}$ precursors (including $CD5^+$ B cells) generate antibody forming and not memory cells. In view of the observations quoted above that activated B cells generally are excluded from follicles, one can speculate that $CD24^{lo}$ B cells differ in this respect from other B cells. Thus, it is

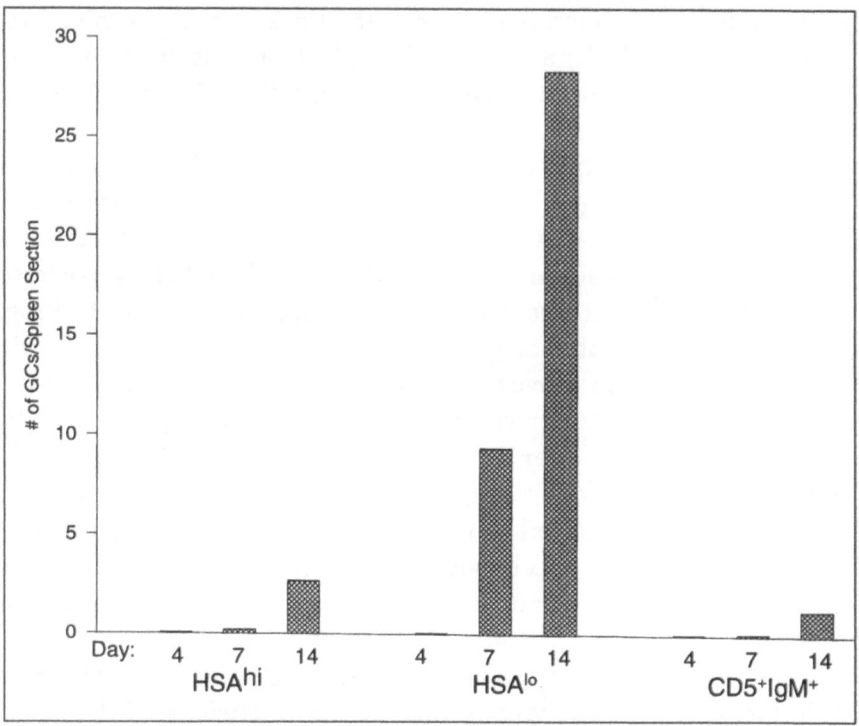

Fig. 1.5. Frequency of GC formation in C.B17-SCID recipients by B cell subpopulations isolated from normal BALB/c mice. Limited numbers of B cells of a given subset (isolated by preparative cell sorting of I-a$^+$CD24 (HSA)^{hi}, I-a$^+$CD24^{lo}, and CD5$^+$sIgM$^+$ cells after lysing of T cells) were transferred i.v. together with 2x10^6 hemocyanin (Hy) primed T cells into C.B17-SCID recipients, that were immunized with 100 µg dinitrophenylated (DNP)-Hy i.v. and 2 x 10^7 *B. pertussis* i.p. On the indicated days (4, 7, 14), recipients were killed, their spleens removed, and the number of GCs per spleen section was determined after staining for PNA binding or Ig. The results are expressed as the #s of GCs/ spleen section found per 10^5 B cells transferred. Data taken from Linton et al.[29]

possible that the circulating (transferred) CD24^{hi} cells, in response to antigen, T cells and their cytokine products within the periarteriolar sheaths, terminally differentiate to antibody forming cells and never proceed to the follicles. The CD24^{lo} cells, on the other hand, may respond to antigen only after they have already entered a follicle or else migrate to a follicle after activation, where they then proliferate to form GCs and ultimately differentiate into memory B cells. A way in which the high expression of CD24 itself might influence homing of a B cell subset is suggested by the observation that CD24 binding modulates the B cell surface VLA-4 interaction with its ligand, VCAM-1.[106] This interaction may be of particular importance for the homing of GC cell precursors and have a negative effect on such homing by detaining B cells from their path to follicles. However, it should also be noted that long-lived B cells belong to the CD24[lo] subset, while HSA[hi] B cells are short-lived and less mature.[107]

Phenotype of GC T Cells

Although GCs do occur in spleens of some nu/nu mice,[108] depending on the number of T cells present in the mice,[109] antigens such as *B. abortus*, which can induce T-independent primary responses and therefore do induce normal primary responses in athymic mice, fail to induce GCs in the draining lymph nodes of nu/nu mice, whereas they induce strong GC formation in the draining lymph nodes of euthymic littermate controls.[108,110] Passively transferred T cells reconstitute both GC formation in lymph nodes (Fig. 1.6) and secondary (IgG) Ab production to *B. abortus* in nu/nu mice.[108] More recent observations have confirmed that T cells are absolutely required for the reconstitution of GC formation by B cells in SCID[29] and irradiated mice.[111] In view of this need for T cells in GC formation, it is likely that the T cells that are locally present, primarily in the apical light zone of GCs, are important in this respect. The number of these cells is low in primary follicles, but increases during GC formation. Indeed, recent studies with specific antigens in mice have shown that the T cells within GCs usually express the $\alpha\beta$ TCR specific for an epitope on the antigen that was used to induce the GCs.[112,113] Quantitation of T cells bearing the relevant Vβ, combined with the study of BUDR uptake in vivo throughout GC formation, shows that these antigen-specific T cells proliferate within the GCs and follicular B cell areas.[114] These findings are discussed in much more detail in chapter 3.

The surface characteristics of GC T cells in man are CD3[+], CD4[+], CD45Ro[+] (CD45RA[-]), CD16[-], CD5[+], and a significant percentage of them are also CD57(leu7)[+].[27,115-118] The presence of this NK cell marker is of particular interest, since CD57 bearing cells in the peripheral blood are usually CD8[+] (T) or CD16[+] (NK). Under normal conditions there are very few CD8[+] T cells in GCs. Resting CD4[+] T cells do not enter B cell follicles or GCs on their normal recirculation pathway. However, as shown in studies using transferred TCR transgenic CD4[+] T cells, activated T cells, particularly when stimulated by antigen + an adjuvant such as bacterial endotoxin, do enter follicles.[119,120] Accordingly, the T cells found in GCs appear to be in a partially activated state: they display CD69, a phenotypic marker of early activation, but not IL-2R (CD25), CD71 (transferrin R) or HLA-class II.[121]

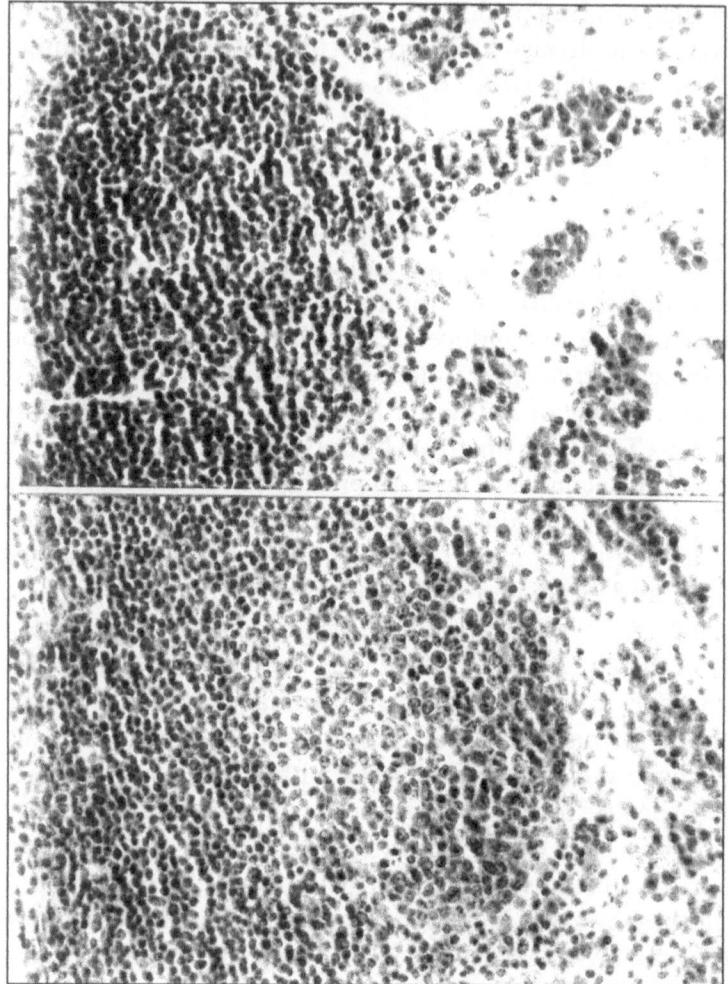

Fig. 1.6. Reconstitution of GC formation in lymph nodes from athymic mice by injection of T cells. 10^7 thymus cells were injected i.p. within the first two weeks after birth and the tissues were examined 1-2 months later. Mice were injected with antigen (killed *B. abortus* vaccine) in the front footpads 10 days before the brachial (draining) lymph nodes, shown here, were taken. Note the total absence of GCs, but presence of normal primary B cell follicles in the cortex and of plasma cells in the medullary cords in the mouse which did not receive thymus cells (top panel). In contrast, the mouse which had received thymus cells produced the typical GCs, of which one is shown here (bottom panel), that are also induced in normal mice by immunization with *B. abortus*, even though the paracortex was still relatively devoid of T cells. Illustration of observations reported on previously by Jacobson et al.[108]   See color figure, page 238.

Activated T cells of memory phenotype, i.e., CD45RO$^+$ T cells such as are present in GCs, are frequently also CD70$^+$(=CD27L).[122] No report on the presence of this marker on GC T cells has appeared to our knowledge, but the fact that CD27 occurs on a subpopulation of memory B cells[123] suggests the possible importance of this ligand pair in GC function.

Murine and rat GC T cells exhibit similar characteristics,[26,124,125] but no report on the presence of NK markers on these T cells in rodents has appeared to our knowledge. Both CD40L and CTLA-4 expression has been found on CD4$^+$ GC T cells in humans[126,127] and in mice (Table 1.1a[56]). In CD40L deficient individuals, GCs are extremely small or absent, but normal numbers of CD57$^+$ CD4$^+$ T cells are present in their B cell follicles.[128] This is perhaps in agreement with the observation that CD57$^+$ T cells do not express CD40L on activation, whereas CD57$^-$ T cells do.[129] In class II MHC$^{-/-}$ mice, in which the remaining few CD4 T cells (a significant percentage of which bear NK1.1[130]) in peripheral lymphoid tissues are confined to B cell follicles, no GCs are produced in response to immunization, probably because class II restricted recognition of antigen is usually required somewhere on the pathway to GC production.[131] Although most of the evidence concerning antigen specific T cells comes from studies in mice, whereas the observations on NK-like T cells come from studies on human GCs, the findings suggest that there are two different kinds of CD4$^+$ T cells in GCs: (a) cells that are specific for the GC-inducing antigen, which exhibit CD40L; and (b) NK-like T cells which resemble the so-called natural T cells, a subset of T cells that in both the mouse and in man have specificity for CD1d and exhibit invariant Vα chains in their TCR, Vα24 in humans and Vα14 in mice, as well as a restricted Vβ repertoire.[132-137] It is not yet clear which antigen(s) govern the development of these natural T cells and what their role, if any, is in GC formation. Studies on β2-microglobulin$^{-/-}$ mice show that these T cells are not essential for Th2 helper cell development.[138] Moreover, CD1$^{-/-}$ mice, which lack NK1.1$^+$ T cells,[134] exhibit normal IgE production and undiminished GC formation in draining lymph nodes in response to immunization with trinitrophenylated hemocyanin[139] (Simmons WJ, Thorbecke GJ and Grusby MJ, unpublished observations).

In the absence of T cell stimulation, such as during the immune response to certain pure polysaccharide antigens,[140,141] GC formation generally does not occur in lymph nodes draining the site of antigen injection. It is of interest, however, that GC formation has occasionally been reported in response to pure polysaccharide antigens, such as to 1→3 dextran preparations[142] and to the pneumococcal phosphatidyl choline determinant.[143] These GCs were recognized in tissue sections by staining with anti-idiotypic Ab or by binding of the antigen. It seems unlikely that these GCs are formed in a T cell independent fashion. More likely is that T cells recognizing the idiotype of this Ab, rather than the polysaccharide antigen itself, are involved in the GC induction during these immune responses. It has previously been observed that anti-Id responses accompanying antibody responses to T-independent antigens are themselves T-dependent and absent in nu/nu mice.[144] Another possibility is that the T cells involved in the production of GCs to nonprotein antigens belong to other T cell subsets that recognize CD1 or CD1 associated molecules rather than class II MHC, particularly as CD1 is expressed on B cells[135,145] and may therefore be involved in the presentation of nonpeptide (glycolipid) antigen determinants. Moreover, when $\alpha\beta$ T cells are lacking and mice are repeatedly stimulated with parasitic antigens, $\gamma\delta$ TCR-bearing T cells are found associated with the rare GCs found in such mice.[146] It is known that $\gamma\delta$ T cells recognize antigens other than typical class I or II associated peptides, such as isopentenyl pyrophosphate and related prenyl pyrophosphate derivatives.[147,148] It should be noted that although $\gamma\delta$ T cells, as well as some CD8$^+$ T cells, are capable of expressing CD40L and of inducing Ig isotype switching in B cells,[149,150] the T cell help for GC formation appears much more effective in CD4 T cells.

Role of Antigen in GC Induction and Maintenance

While stimulation with exogenous antigens does not appear to play a role in the generation of primary B cells in the bone marrow of mammals or in the bursa of Fabricius of chickens, antigenic stimulation seems to be of prime importance in the generation of GCs in secondary lymphoid organs. Germfree chickens, mice and rats lack GCs, particularly in their gut-associated lymphoid tissue.[151-153] During ontogeny in these species, the first appearance of GCs in the gut-associated lymphoid tissue comes only after the bacterial flora has established itself.[154] Nonspecific augmenting factors, such as postnatal transfer of maternal immunity[155] and weekly injections of oligomeric IgD (which induces IgD-receptors on CD4 T cells[97]) accelerate but do not initiate the development of GCs in the lymphoid tissues of neonatal mice.

In the sheep embryo, B cell proliferation occurs in gut-associated lymphoid tissue that consists of tightly packed lymphoid B cell follicles which contain predominantly sIgM$^+$ cells and are intimately associated with epithelium; they develop in a totally antigen as well as thymus independent fashion.[156-158] In contrast to GCs, the blast cells that proliferate are located in the peripheral zone of these follicles, resembling the structure of the bursa of Fabricius in the chicken.[157] The injection of antigens into the lumen of isolated ileum segments in sheep embryos causes the development of typical GCs in the regional mesenteric lymph nodes and of plasma cells in the lamina propria and in lymph nodes.[158] The follicles in segments of the ileum that are devoid of antigen disappear by 3-4 months after birth of the sheep, whereas the follicles in the normal ileum and jejunum do not involute until the sheep are around 15 months of age. Therefore, it seems that in sheep also, the formation of GCs and plasma cells is antigen dependent.

GC formation does not appear to be induced by stimuli other than immunogenic ones. In experiments specifically directed at this question it was found that, in rabbits or mice, nonspecific adjuvants such as alum or silica alone fail to induce GCs.[159] Neither do protein antigens to which tolerance has first been induced, even when they are injected with adjuvants.[160,161] In view of the need for antigen specific T cells as helpers for GC formation, it is not immediately evident whether the defective GC formation in tolerant animals is due to tolerance at the T or B cell level. Most of the recent evidence suggests that it is primarily due to T cell tolerance,[6] although the tendency of tolerant B cells not to enter into follicles and to get arrested in the outer T cell zones instead, is very dependent on their prior exposure to specific antigen, the degree of crosslinking of their sIg,[162] and the nature of the antigen.[163]

The antigen localized on FDCs is thought to be of great importance for GC induction (reviewed in refs. 164-167). FDCs are able to fix immune complexes on their cell surface by virtue of their FcγR (CD16 and CD32), FcεRII (CD23) and receptors for C1 (CD35), C2 (CD21; a specific isoform that comprises one more exon than the CD21 on B cells) and C3 (CD11b).[168-171] Immune complexes containing the antigen in undegraded form are maintained on the surfaces of FDCs for prolonged periods after antigen injection.[172-174] This property appears critical for the long term maintenance of specific IgG Ab levels (reviewed in Tew et al[165] and see chapter 2). Any antigen capable of activating C3 or contained in immune complexes, including intact viral particles,[175] may localize on FDC dendrites. Mannose receptors on dendritic cells are thought to be involved in the binding and transport of carbohydrate antigens to FDCs.[176] It has also been reported that elimination of marginal zone B cells inhibits the trapping of immune complexes on FDCs, suggesting that B cells carry immune complexes to the FDCs.[177] Heinen et al[178] have shown that B cells coated with immune complexes can transfer them to FDCs in vitro. Several viruses have also been found as intact virions ensnared within the FDC dendritic network.[175,179-183]

In ontogeny in the rat, the first production of GCs on immunization is correlated with the appearance of FDCs.[34,184] Antigen localized on the surface of FDC is thought to play a role in the induction of GCs through the formation/release of immune complex coated bead-like structures, called "iccosomes" by Szakal et al,[185] that could in conjunction with adhesion molecules on the FDC surface serve to stimulate the antigen specific B cells directly. Alternatively, antigen could be taken up by GC B cells from FDCs via sIg receptors, endocytosed, processed and presented with MHC class II to helper T cells in their vicinity with a resulting T-B cell interaction that leads to B cell proliferation.[186-188] It is thought that the inducing antigen persists (in immune complexes) on the surface of FDCs and is presented by activated B cells to the T cells within the GC. However, the possibility has not been ruled out that the direct recognition of surface antigens on the GC B cells by T cells could not also be involved in the maintenance and continued stimulation of GC formation. If this is true, GC formation initiated by a conventional antigen could be perpetuated by T-B cell interaction perhaps even in the absence of the initiating antigen. The possible role of superantigens (SAgs) expressed on the B cell surface in the induction of such continued stimulation and the resulting hyperplasia of the B cells and GC derived lymphoma formation is discussed in chapter 5 (see also Ponzio et al[189]).

To date, attempts to induce GCs with bacterial SAgs in normal mice have not been successful in our hands (Lakhman J, Thorbecke GJ, unpublished observations). However, modest success has been described in this regard with viral SAgs, which were found to cause GCs in draining lymph nodes associated with the relevant Vβ-restricted T cell expansion.[190] It is possible that SAgs capable of stimulating T cells expressing only β chain or β chain with an α chain precursor in their TCRs are responsible for the GC formation in

Table 1.2a. Germinal centers in mice engineered for differences in individual B or T cell or FDC surface antigen expression

Genetic Marker*	Major Immune Function Change	Germinal Centers	Refs.
Fas$^{-/-}$	Autoimmune Abs↑; B1↑	Normal	67, 543
CD40$^{-/-}$	Absence of Ig Switching; Prim. + Sec. Ab Resp. ↓ Lack of T-cell Priming	↓↓ ; Particularly in TD Resp.; CD40-Fc-γ1+Ag Partially Restores	292,544, 545 297
CD19$^{-/-}$	TI Ab Resp.↑; TD Ab Resp.↓	↓↓↓ (Follicles Normal)	247,546
CD19 Tg	TD Ab Resp.↑;TI Ab Resp.↓	Background Normal; But Resp. to Ag↓↓	247,547
CD81$^{-/-}$	TI Ab. Resp.↑;CD19↓↓;B1↓↓	Probably Normal	548
CD22$^{-/-}$	B cells Activated; Ca^{2+} Resp.↑↑; Long-lived B↓; TI-2 Ab Resp.↓; IgG3↓	Normal	549,550, 551
IgD$^{-/-}$	None	Normal	102
I-a$^{-/-}$	CD4 T↓↓, TD Ab Resp.↓↓↓	Absent	131
CD-1$^{-/-}$	Absence of NK-like T cells	Normal**	139
CD40L$^{-/-}$	Absence of Ig Switching	Absent	552,291
CD43 Tg	B cells↑; TI & TDAb.Resp.↓↓ Defective T-B interaction	↓↓ (Follicles Normal)	553,554
CD28$^{-/-}$	TD Ab Resp↓; Ig Switch↓; Ab Maturation Absent	Absent, also in PP (Follicles Normal)	555
B7-(1+2)$^{-/-}$	IgG1 + IgG2a Resp.↓↓↓	Absent	535
B7-2$^{-/-}$	Resp. to Ag iv (No Adj)↓↓	↓↓ in Resp. to Ag i.v.	535
CTLA-4-Ig Tg	TD Ab Resp↓; Ig Switch↓ Ab maturation poor	↓↓↓ (but Present in PP)	556,557
TCR-α$^{-/-}$	Auto-Abs; IL-2↓; IL-4↑; IFN-γ↑; γδ T cells↑↑	Some, but No Resp. to Immunization	146,558, 476
TCR-β$^{-/-}$	TD Ab Resp.↓↓	Absent (except with Parasite Infections)	559,560
CD23$^{-/-}$	IgE Ab Resp.↑↑	Normal	561
CD32/ FcγRII$^{-/-}$	Prim. + Sec. Ab Resp.↑↑; PCA↑↑	Normal	562
CD21/35$^{-/-}$ (Cr2&1)	TD Ab Resp↓↓; B1a ↓; TI Ab Resp Normal	↓↓ in # + size (due to B cell Defect)	482,483
P- + E- Selectin$^{-/-}$	Increased Hematopoiesis; Leukocytosis; Infections	Normal	563
L-selectin$^{-/-}$	LN: B Foll.↓; IgM↓; Resp. to subc.Ag: Prim.↓, Sec.↑	↑↑ in SPL + LNs	564

See footnotes following Table 1.2c.

Table 1.2b. Germinal centers in cytokine, cytokine receptor and complement component knockout mice

Genetic Marker*	Major Immune Function Change	Germinal Centers	Refs.
LT$\alpha^{-/-}$	Absence of LNs + PP; FDC↓↓↓; Abnormal PALS Organization ("Ringing")	Absent (except in rare LN); Maturation of Ab (at high Ag dose)	487,488, 565
LT$\beta^{-/-}$	Absence of most LNs + PP FDC↓↓↓; PALS "Ringing"	Absent in SPL and PP; Some in LNs	494,498
TNF-$\alpha^{-/-}$ or TNF-RI$^{-/-}$	PALS "Ringing"; Also no Follicles in PP or LN; FDC↓↓↓; MAdCAM-1 in marginal zone ↓↓	Absent in SPL, PP or LN	496,495 492,493
TNF-RII$^{-/-}$	Normal	Normal	493
LTβ-R-Fc Tg	PALS "Ringing"; Also few if any PP; LN Normal	Absent in SPL; Some in PP and LN	503
LT Tg	Ectopic Lymphoid Tissue in Targeted Organ	Normal; Also Present in Ectopic Lymph. Tissue	504
BLR1$^{-/-}$	Defective Homing of B cells into Follicles of SPL + PP; Some LNs absent	Absent in SPL + PP; Normal in LNs	270
IL-4$^{-/-}$	Mucosal Immune Resp.↓; Low IgG1 and IgE	↓↓ in PP (strains differ) Normal in LN + SPL	352,§
IL-2$^{-/-}$	Lymphoprolifer. Disease; Auto-Abs/Immunity	Hyperplastic; Anti-CD40L Prevents	566
IFN-γR$^{-/-}$	Anti-viral Abs ↓; Resp. to BCG↓	Normal	567,568
IL-5R$^{-/-}$	B-1 cells↓↓; Eos & Eos Resp.↓↓; IgM + IgG3↓	Normal	569,‡
IL-5$^{-/-}$	B-1 cells↓↓; Eos↓↓	Normal	354
IL-6$^{-/-}$	IgG Ab Resp.↓↓	↓; Abnormal Structure	354
IL-6 Tg	Plasmocytosis	Hyperplastic	355
C3$^{-/-}$ or C4$^{-/-}$	TD Prim.+ Sec. Ab Resp.↓; TI Ab Resp OK; Ig Switch↓↓	↓↓ (Defect overcome by High Ag Dose)	478

See footnotes following Table 1.2c.

Table 1.2c. Germinal centers in intracellular signal mediator knockout mice

Genetic Marker*	Major Immune Function Change	Germinal Centers	Refs.
OCA-B/ OBF-1$^{-/-}$	B1↑; Mature B ↓↓; B cell Resp. to Ag↓↓↓	Absent	570,571, 572
Eµ-Bcl-2 Tg	Naïve +Memory B cells↑↑; Autoimmune Abs↑	↑ in Resp. to Ag; Corticoid Resistance ↑	453,452, 437
Bcl-2$^{-/-}$	B + T cells↓↓↓; Ly Lifespan↓	↓↓	229
Bak$^{-/-}$	Hyperplastic Thymus + B cells	↑**	573,§
BCL-6$^{-/-}$	Th2 Cytokines ↑↑	Absent	526
LSIRF/IRF4$^{-/-}$	Mature B cells ↓↓; Ig↓↓↓; Plasma cells↓↓↓	↓↓↓	574
NF-IL-6$^{-/-}$	Plasma cells ↑↑; IL6↑↑; Prevented by IL6$^{-/-}$	↑↑ (Resembles Castleman's Disease)	575,576
p50/NF-κB1$^{-/-}$	B cell Prolif.↓, Ig Resp.↓ Switch to IgG3 + IgE↓	Normal	577,520
RelB$^{-/-}$	Thymic Stroma DC↓↓; B cell Prolif.↓, Maturation Normal; T-cell Infiltrates	Normal	517,518, 577
RelB$^{-/-}$ + p50$^{-/-}$	Generalized Inflammation Thymus↓↓↓; B cells↓↓	↓	520
Bcl-3$^{-/-}$	TD Ab Resp.↓; LN size↓ B Follicles↓↓↓;FDC↓↓↓ B cell Ringing of PALS	↓↓↓ (not Absent)	513
p52/NF-κB2$^{-/-}$	TI-1 Resp↓↓; TD Resp.↓ B-cell Foll↓↓↓; FDC↓↓↓	↓↓ ↓ (not Absent)	512
STAT6$^{-/-}$	Th2 Resp.↓↓↓; IgE↓↓	Normal**	578
STAT4$^{-/-}$	IL-12-induced Th1 Resp↓↓ Th2 ↑↑; NK↓	Normal**	578
A-Myb Tg	B cells↑↑; SPL + LNs↑↑	↑↑; Polyclonal Hyperplasia	579
SHP-1$^{-/-}$/ Motheaten	B cells Spont. Activated; sIgM↓; Autoimmune Abs	Normal***	580,581
Lyn$^{-/-}$	CD40 Signaling↓; Lymphoid Hyperplasia; Mature B↓; B1↑↑; Auto Ab↑↑; IgM↑	↓↓↓	527,582, 583
Btk$^{-/-}$(XID)	% Long-lived B cells↓; TI-2 Ab Response↓	↓; Memory B cells ↓ Somatic Mutation OK	544,421
Btk$^{-/-}$ + CD40$^{-/-}$	% Long-lived B cells ↓↓; Most Ab Responses ↓↓↓	↓↓↓	544,584

See footnotes for Tables 1.2a,b,c on following page.

Table 1.2a,b,c footnotes

*   Tg designates overexpression of transgene; -/- designates knockout gene.
    Ag = antigen; Ab = antibody; CFA = Complete Freund's Adjuvants; LN = Lymph nodes;
    Spl = spleen; Resp. = response; Prim. = primary; Sec. = secondary; TD = Thymus dependent;
    TI = thymus independent.
** Simmons WJ, Thorbecke GJ and Grusby MJ, unpublished observations.
‡ Takatsu K, personal communication.
§ Unpublished observations on mice donated by Drs. S.J. Korsmeyer and L. Rizzo.
***It is of interest that GC cells were recently reported to express much lower levels of the
    phosphotyrosine phosphatase, SHP-1, a negative regulator of B cell activation, than mantle
    zone B cells, suggesting that they have a low threshold for activation.[584a]
    Knockout mice lacking B cells altogether, such as Ikaros[-/-, 528] $\mu^{-/-,585}$ and rag-1 or rag-2[-/- 586]
    mice or mice greatly deficient in peripheral B cells, such as γ_c chain[-/- 587] and jak3[-/- 588,589] mice,
    were omitted from these tables. Unfortunately, mice deficient in β_c, the common cytokine
    receptor chain, in which responses to IL-5 and GM-CSF are greatly reduced,[590,591] could not be
    included, because no reports on the status of GC formation in these mice are available to our
    knowledge.

TCRα[-/-] mice (see Table 1.2a and under genetic models). An important factor determining the ability of antigens to induce GCs may be
the duration of the T cell activation induced in vivo. In this respect
SAgs are inferior to regular antigens, as even when injected with
complete Freund's adjuvant, SAgs induce anergy in the responding
T cells within 10 days, whereas regular antigens do not.[191]

Trapping of Antigen Specific B Cells from the Circulation by Immune Complexes on Follicular Dendritic Cells

An additional role of the long-term FDC bound antigen present
in follicles from lymph nodes draining the site of primary antigen
injection appears to involve the capture of recirculating memory B
cells with high affinity for the antigen in those follicles. As late as 4-6
months after priming with a hapten conjugate (without adjuvants)
in a single footpad in mice, B cell memory is found primarily in cells
from the draining lymph node and can no longer be transferred with
cells from the contralateral node[192] (Fig. 1.7), suggesting a persistent
local component in the B cell memory, also noted by others.[193] Similarly, long-term memory in rabbit lymph nodes draining a footpad
injection of a very low dose of diphtheria toxoid (without adjuvants),
as detected by the ability to elicit a secondary Ab response from the
tissue in vitro, remains higher in the draining than in the contralateral lymph node for more than one year after priming.[194] This phenomenon of local memory was analyzed in more detail, taking advantage of the adoptive memory B cell approach.[195-197] It was found
that in recipients which themselves are inhibited from responding,

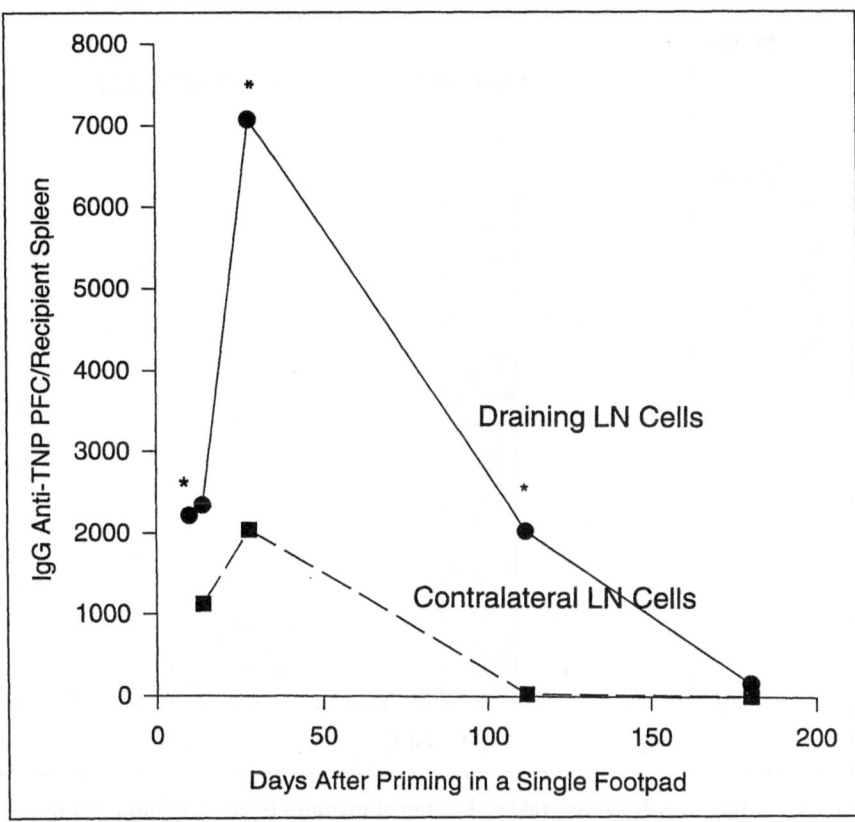

Fig. 1.7. Immunological B cell memory in draining and contralateral lymph nodes of LAF1 mice after priming with TNP-hemocyanin in a single footpad. Mice were injected with 15 µg TNP-hemocyanin in one front footpad and at various intervals after this 5 x 10^6 cells from the draining and from the contralateral brachial lymph nodes were transferred i.v., together with 10^7 bovine Ig (BGG) primed purified T cells and 100 µg TNP-BGG, into 620R whole body γ-irradiated syngeneic recipients. Spleen cells from recipient mice were assayed for numbers of anti-TNP secreting IgG secreting cells ("indirect PFC") 7 days after cell transfer. Note that the cells from draining nodes transferred significantly higher IgG antibody responses to TNP at all intervals, from 10 days to 26 weeks, after priming ($p< 0.05$ at 2 and 26 weeks, $p< 0.01$ at 4 weeks and <0.005 at 16 weeks) (data taken from Baine et al[192]). The asterisks indicate that similar experiments were performed at these times after priming in which cells from the draining lymph node binding to PNA-coated dishes (i.e., GC cells) were separated from other purified lymph node B cells and transferred separately. Care was taken in those experiments to ensure the presence of GCs in the donors even at the later time points after priming with trinitrophenylated (TNP)-hemocyanin (KLH) by injecting B. abortus antigen into the footpad 10 days prior to transfer. This did of course induce GCs, but these GC B cells (isolated at 1-6 months after priming with TNP-KLH) were not effective at transferring memory responses to TNP, whereas the PNA⁻ B cells were. The only time at which PNA⁺ B cells from the draining lymph node were much more effective than PNA⁻ B cells at transferring anti-TNP memory was at 10 days after priming with TNP-KLH, i.e., at the time of maximal GC cell proliferation induced by TNP-KLH itself. Cells from the contralateral lymph node which transferred memory for the anti-TNP response were always PNA⁻ (Coico et al[25]).

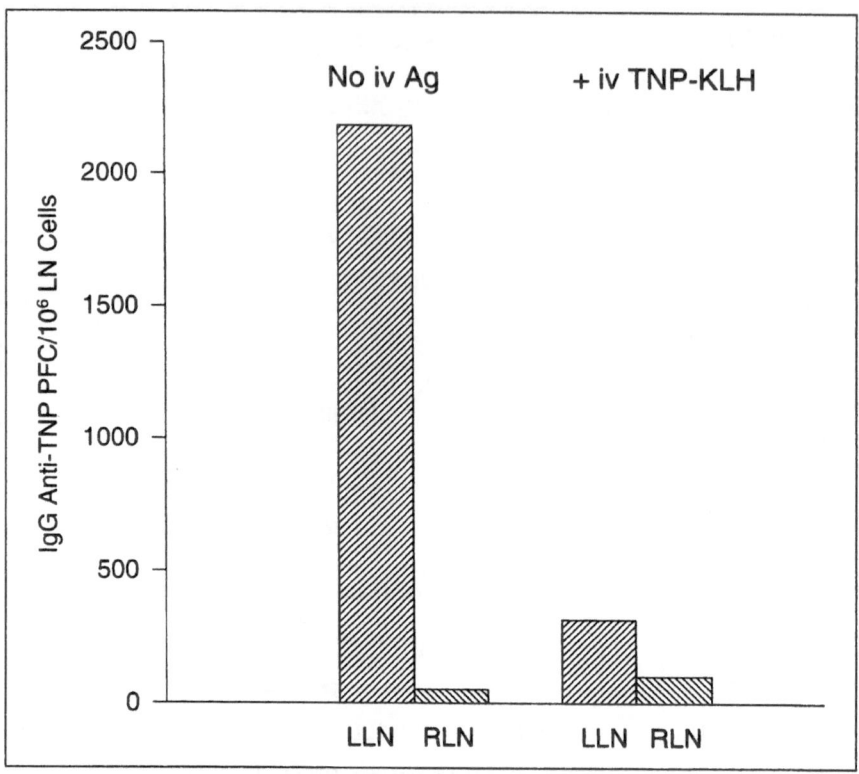

Fig. 1.8. Antigen specific unilateral localization of memory B cells is inhibited by intra-venous administration of the antigen. Recipient LAF1 mice were pretreated with cyclo-phosphamide (250 mg/kg body weight) 1 day before injection of 10 µg TNP-hemocya-nin in the L front footpad and 10 µg arsenylated (ARS)-KLH in the R front footpad. They received 4 x 10^7 TNP-BGG primed spleen cells with or without 100 µg TNP-KLH i.v. and were challenged in both front feet with 80 µg of TNP-*B. abortus* 5 days later. Indirect PFC/10^6 lymph node cells were determined 4 days after challenge. The dif-ference between the numbers of PFC in the L and R lymph nodes was significant (p<0.025) without, but not with, the i.v. injection of TNP-KLH into the recipients. These results strongly suggest an antigen (TNP)-specific localization of the memory B cells in the draining as compared to the contralateral lymph node in the absence of any im-mune response in the host (no PFC detected without cell transfer). (Data taken from Baine et al[195]).

but which harbor hapten-containing immune complexes in their lymph nodes, hapten-specific memory B cells preferentially localize in lymph nodes draining the site of injection of the same hapten(-conjugate) as compared to lymph nodes draining the site of injection of a different hapten. Intravenous injection of the same hapten (on a different protein carrier) interferes with the unilateral localization of the hapten specific B cells[195-197] (Fig. 1.8). The results of these experiments, also addressing the ability of these adoptive memory B cells to proliferate and to be challenged into Ab responses, suggest that antigen (immune complexes), localized on FDC in follicles of the draining lymph node long after the disappearance of the initially induced GCs, is capable of retaining antigen-specific resting memory B cells from their recirculation pathway without inducing them to proliferate or become Ab secreting cells unless another footpad challenge with antigen activates T cell help in that lymph node. The evidence that a significant proportion of memory B cells is still present as long-lived, nondividing cells several months after priming[198] is in agreement with this interpretation. It seems possible that being retained in lymph node follicles through antigen fixed on FDCs prolongs the survival of these memory cells.

Functions of GCs and the Cellular Interactions that Contribute to These Functions

The functions of GCs differ somewhat in different species. In most mammals, GCs are sites where the generation of memory B cells occurs, accompanied by Ig gene V region somatic hypermutation.[199] However, GCs are also used as sites for primary B cell Ig gene diversification by some species, particularly the rabbit.[24,200,201] In addition, somatic hypermutation, although a hallmark of GCs in most mammals, is used for the diversification of the Ig repertoire in lymphoid follicles without GCs in the sheep embryo. The discussion of these subjects in this chapter serves as an introduction to the more specialized treatise of somatic hypermutation given in chapter 4.

Generation of Memory B Lymphocytes

The observation that GC formation comes after the peak of antibody production, both in the primary and in the secondary response, suggested the idea that GCs might represent sites where memory cells were produced.[8,202] Various approaches were adopted in attempts to obtain evidence for this possibility. It was reasoned that these temporal relationships should allow interference with memory cell generation in parallel with inhibition of GC formation, without an appreciable effect on the primary antibody response. Indeed, when drugs, such as colchicine, prednisolone and cyclophosphamide, or γ-irradiation were applied during the peak of GC proliferation, memory for the secondary antibody response was found to be significantly delayed without much of an effect on the primary antibody response.[203-206] More recently, similar results have been obtained with mAbs to CD40 or B7-2, which also disrupt GCs when applied at the peak of GC formation.[207] It should also be noted that, for instance in the case of γ-irradiation or prednisolone, the ability to produce a secondary Ab response usually recovers in parallel with the histological recovery of GCs.[205,208] This recovery of GCs is probably due to the survival of a few GC blasts in the presence of persisting immune complexes on the FDCs. Recovery of memory for the secondary Ab response is much slower or not detected, if γ-irradiation is applied several weeks after priming, i.e., after GC cell proliferation induced by the primary injection of antigen has waned.[208-210] In contrast, if GC formation is accelerated, as after priming with immune complexes rather than with free antigen, B cell memory also develops earlier.[211,212]

Another approach showed that secondary responses in vitro were detectable in the white pulp of the spleen at peak GC formation and not in the red pulp at that time or at any time before that.[8] Reasoning that GCs must be producing small recirculating memory lymphocytes, Wakefield et al[213,214] studied the fate of the progeny of (³H)thymidine-labeled GC-containing white pulp in syngeneic recipient mice in the absence of antigen and showed them to become small lymphocytes rather than plasma cells. Exciting as this was at the time, a lot of this work, done with isolated white pulp taken at peak GC formation, became very difficult to interpret when it became clear that there were not only two different types of small lymphocytes, T and B, but that immunological memory was a property of both these populations.[215,216]

Subsequent studies, involving cell transfers and in situ observations, confirmed that expansion of B cell memory correlates temporally with GC formation,[25,192,211,217] whereas T cell memory expands much earlier after antigen injection.[218] However, a more specific marker for GC cells was needed to allow isolation of GC cells. The PNA binding property of GC B cells[23] provided such an approach.[25] In the presence of excess carrier-primed T cells, adoptive memory responses can be transferred with the PNA⁺ fraction of B cells isolated from lymph nodes draining the site of antigen injection in mice as early as 7-10 days after priming with hapten conjugate in the front feet. PNA⁻ B cells do not transfer a memory response until day 14 and at that time these (PNA⁻) memory B cells can already also be detected in the contralateral lymph node[25] (see also Fig. 1.7). It was concluded from this work that memory B cell generation occurs in GCs and that the PNA⁺ B cells rapidly mature into PNA⁻ memory B cells that inter the circulation and thereby become disseminated over the lymphoid tissues. It should be noted that at later times after priming, in both the rabbit and in mice, memory B cells are readily detected in the bone marrow as well as in lymphoid tissue.[7,219]

The recent advances made by Liu and coworkers[47] in the isolation of human tonsillar B cell subsets include the clean separation of CD38⁻IgD⁻ memory B cells from CD38⁺ GC cells and IgD⁺ naive B cells (see Table 1.1b). As a result, the properties of memory B cells have now been compared with those of naive B cells using in vitro culture with CD40L-transfected fibroblasts, IL-2 and IL-10, i.e., the cells are triggered in an Ig-independent fashion. The long suspected difference between naive and memory B cells in their ability to differentiate into plasma cells, which was previously ascribed to differences in avidity for the antigen and/or density of surface Ig, was clearly shown in these experiments. Memory B cells give rise to 5- to 8-fold more plasma cells than do naive B cells, whereas naive B cells produce relatively more nondifferentiated blast cells.[220] This observation may explain the extensive extrafollicular, Ab-containing plasma blast proliferation seen within 2 days after a secondary injection of antigen, as opposed to the much smaller foci seen on days 4-5 after a primary injection.[11] Similarly, sIgA⁺ memory B cells from Peyer's patches rapidly differentiate into plasma cells on exposure to IL-6 in vitro, whereas PNA⁺ sIgA⁺ GC blasts do not.[221] It is also of interest that Toellner et al[12] have identified memory B cells, that have recently been challenged with antigen, as a major cell type in which isotype switching occurs, exhibiting switch transcripts within 12-24 hours after antigen challenge, whereas they can be found only by day 4 after the primary response and at much lower levels. These investigators find the continued production of these switch transcripts throughout the period of GC formation, suggesting that switching also occurs in GCs, as clearly shown before,[31-33] but in a relatively small percentage of the B cells.

Cellular Interactions Important in the Induction of GCs
The formation of GCs is likely to be dependent on the interactions between T and B cells, between FDC and B cells, and possibly also among B cells themselves. There is no clear distinction between those interactions which lead to the local proliferation of B cell blasts and those which primarily prevent apoptosis of GC B cells. In fact, some of the cell interactions probably serve both functions at once. GC B cells show a high rate of spontaneous apoptosis when placed in tissue culture,[222,223] particularly among cells bearing CD77⁺, the

Table 1.3. Cytokines produced by germinal center cells

Cell Type	Cytokine/Chemokine	Reference
GC T	IL-4, IL-10, TNF-α and IFN-γ	Butch et al[328]
Interdigitating Dendritic	DC-CK1	Adama et al[269]
Follicular Dendritic	IL-6	Orscheschek et al[243]
	HGF/Scatter Factor	Van der Voort et al[267]
	NGF	Strobach et al[253]
GC B	G-CSF	Corcione et al[329]
	IL-10, TNF-α	Burdin et al[262]
(anti-CD40+ SAC) Activated Follicular Mantle/Memory B	IL-6, IL-10, TNF-α	Burdin et al[262]

expression of which appears related to the relative activities of different glycosyl transferases.[38,224] Interaction with FDCs[225-227] and/or activated T cells expressing CD40L[228,229] protects against this apoptosis. As is summarized in Table 1.3, a number of Abs to and/or ligands for surface molecules of GC B cells can mediate this effect. Anti-Ig and anti-CD40 are listed among these Abs with protective activity for at least part of the GC B cells. These Abs in particular are thought to mimic physiological signals provided by surrounding cells within the microenvironment of the GC in vivo. It is reasoned that those cells capable of binding the antigen (or immune complex) on the FDC surface with high affinity, and therefore able to present that antigen to the antigen specific CD40L[+] T cells in the apical light zone of the GC, are protected and thereby positively recruited into the surviving centrocyte and memory B cell populations.

FDC-GC B Cell Interaction

FDCs are nonphagocytic cells characterized by long branching processes which form a delicate three-dimensional network throughout B cell follicles and GCs.[230-232] The enlarged FDC meshwork in GCs that is found after antigen injections is full of connexin43, one of the proteins that make up the hemichannels of functional gap junctions between FDCs and between FDCs and CD19[+] B cells.[233] Most of the connexin43 containing gap junctions are in the light zone,[234] where the FDC envelop maturing B centrocytes and a few CD4[+] T cells. It is not yet known which substances move from cell to cell under normal conditions, but it is likely that this communication between the cells serves a role in the function of GCs.

In the absence of B cells, FDCs have smaller and fewer processes and are considered precursors of FDCs.[235] In SCID mice, few FDCs are found and their reconstitution by cell transfers showed that the combination of B and T cells was more effective than B cells alone.[236] This finding, as well as the observed paucity of FDCs in follicles from CD40L deficient individuals,[128] suggest that T-B cell interaction is stimulating for FDC development. In experiments where rat bone marrow or fetal liver cells were used to reconstitute γ-irradiated SCID mice, FDCs of rat origin were found in some of the recipients, suggesting that at least some of the FDC may be of hematopoietic origin.[237]

As described in chapter 2, the immune complexes kept on the surface of FDCs in the form of "iccosomes" are presented to and endocytosed by B cells. During this interaction the additional surface molecules that strengthen the contact and mediate the protection of the GC B cells from apoptosis include several adhesion molecules. ICAM-1 (CD54), VLA 3, 4, 5 and 6 (CD29/CD49c-f)[46,168,170] and vascular adhesion protein 2 (VAP-2 [CD73][238]) are all expressed on FDCs, as is VCAM-1 (or INCAM-110), a ligand for CD29 (VLA-4 and/or -5[239]). Thus, these cells not only exhibit all the complement and Fc receptors that help them bind antigens and immune complexes, but also many adhesion molecules important for cell interactions. There appears to be a correlation between the molecules important for FDC-B cell interaction and those that mediate protection against apoptosis. These include in particular the VCAM-1 - VLA-4 and the ICAM-1 - LFA-1 interaction pairs of adhesion molecules.[225,240,241] The protection from apoptosis afforded by FDC is reported to involve rapid and irreversible blockade of endonucleases such as nuc-18, and to be quite distinct from the effects of CD40, CD21, or sIg ligation.[227] On the other hand, Abs to CD45, CD44, CD2, ELAM-1 or MHC class I antigens do not inhibit the adherence of FDC and B cells to each other and also fail to influence apoptosis. Anti-CD73 inhibits the aggregation of isolated GC cells, presumably by blocking CD73 on FDCs, as it is not present on most GC B cells.[238] FDC specific markers have been described for both humans (HJ2;[242] Ki-M4[243]) and mice (FDC-M1[167]). No particular function in the interaction with other cells in GCs has as yet been identified for these molecules.

CD21 (CR2) has potentially two ligands on the surface of FDCs that may elucidate the physiological significance of the CD21-mediated protection from apoptosis in the GC: C3 (fixed to immune complexes) and CD23, expressed on FDC in the outer zone of the GC. It has been reported that soluble CD23 in combination with IL-1 provides a differentiation rather than a proliferation signal to the GC B cells that it rescues.[244] Anti-CD19 also decreases GC-cell apoptosis, particularly in synergy with anti-Ig or anti-CD40, and its effect resembles that of IFN-α.[245] In studying the protective effect of anti-CD19, it is important to realize that there are functional and physical associations between CD19 and CD21 on B cell surfaces.[246] CD19 is associated with an important signal transduction pathway that regulates B cell differentiation.[247] It has been shown with Burkitt's lymphoma cells (RAMOS) that targeting of the CD19-associated tyrosine kinase LYN, using Ab to CD19 conjugated to genistein, induces apoptosis.[248] An additional interesting aspect of CD19 is that its extracellular domain has a binding site for CD77 with similarity to the verotoxin B subunits. CD19 and CD77 cocap, as shown on the surface of Daudi cells, and expression of CD19 is influenced by the presence of CD77.[249]

Another important aspect of the interaction between GC B cells and FDCs might be in the production of cytokines and/or chemokines. TNF-α and LT molecules occur as trimers on the surface of cells producing them[250] and activation by anti-CD40 or anti-Ig induces the production of TNF and LT in B cells.[251,252] NGF-R[253,254] and TNF-R (p55)[255] are present on the surface of FDCs. Thus, interaction between FDCs and GC B cells through TNF/LT and their receptors is quite likely to occur in GCs. However, at this time the exact nature of the consequences of such interaction is not yet understood. The absence of FDCs in the few GCs seen in TNF and/or LT$^{-/-}$ mice (see Table 1.2b) suggest that B cells induce FDC development through this interaction (discussed further in the section on knockout mice). Human[256] and murine[257] B cells are also capable of NGF production, but there is as yet no evidence for a role of NGF in FDC-B cell interaction.

A list of the cytokines that have been reported to be produced in FDCs is presented in Table 1.3. The IL-6 containing culture supernatant of the FDC-like cell lines (HK cells, originating from human tonsil) has been reported to costimulate anti-CD40 activated B cells, but the responsible factors were not identified.[258] The reports concerning the identity of cytokines produced by isolated human FDCs are somewhat contradictory. Clark et al[259] find that they produce IL-6, which is upregulated by various cytokines including TNF-α, but they find no evidence for the production of other cytokines, whereas Ryffel et al[255] report the production of TNF-α. CD23 is shed from the FDC surface and produced by IL-4 stimulated B cells; it could be considered a cytokine in that the soluble 12 and 25 kDa glycoproteins derived from CD23 have been found to promote, together with IL-1, the differentiation of a subset of human GC cells to plasma cells.[169,260] The immortalized FDC cell line described by Orscheschek et al[243] produces CD23, IL-1α and some IL-6, but no IL-1β or any other tested cytokine. In Castleman's disease, hyperplastic GCs are found which contain FDCs expressing mRNA for IL-6.[261] This is of particular interest in view of the inability of GC B cells to produce IL-6, even upon activation by CD40L and/or SAC, in contrast to other B cells (see Table 1.3), while CD40L activated GC B cells exhibit increased proliferation in the presence of added IL-6.[262] It is likely that the enhanced proliferation of the GCs in Castleman's disease is related to increased IL-6 production, also because similar hyperplastic GCs are seen in IL-6 transgenic mice, whereas IL-6$^{-/-}$ mice exhibit reduced GC formation (Table 1.2b). A very important recent finding in this respect may also be that FDCs produce IL-7.[263] Since IL-7 influences RAG-1 and RAG-2 expression in the thymus,[264] this IL-7 production in FDC cells might play a role in the reexpression of these proteins in GC B cells.[71,73] It is also known that IL-7 primes CD4 T cells for IL-4 production[265] and may thus influence the Th1/Th2 cell ratio locally within GCs. Unfortunately, studies on IL-7$^{-/-}$ mice report a marked defect in B cell maturation, but do not comment on GC formation.[266] Another influence of FDCs on GC B cells might involve their tendency to adhere to each other and to FDCs themselves. Recently, the c-Met tyrosine kinase receptor for hepatocyte growth factor/scatter factor (HGF/SF) was found to be specifically expressed on centroblasts in the dark zone of GCs, whereas FDC-enriched stromal cells from tonsils were shown to produce HGF/SF.[267] HSF/SF was shown to induce integrin-mediated adhesion of *c-met* cDNA transfected cells (and of Burkitt cells) to VCAM-1 and fibronectin.[267]

Chemokines are probably of foremost importance as chemotactic factors, governing the homing of lymphoid cells. A beginning of our understanding of the mechanism by which these mediators influence the locomotion of cells is from work such as that of Del Pozo et al,[268] which shows the involvement of uropods in lymphocyte recruitment by chemokines and the redistribution of ICAM-1 and -3 to uropods of T cells. Migration through confluent monolayers of endothelial cells is probably also dependent on such uropod formation. The selective responsiveness of lymphocyte subpopulations to chemokines[269] and the production by sessile cells such as interdigitating DCs and FDCs of chemotactic factors could be determining the movement of specific cells into and out of follicles. In this respect it is of interest that defective homing of B cells into follicles and severe reduction in GC formation are observed in mice lacking the chemokine receptor BLR-1.[270] A link between chemokines and TNF/LT is suggested by the finding that chemokine production may be induced in certain cells by crosslinking of the LTβ-R.[271]

IDC-T and/or B Cell Interactions

The presence of dendritic cells other than FDCs in GCs has recently been described.[272] These cells resemble the interdigitating cells from the paracortex of lymph nodes and Langerhans cells more than they do FDCs. Like Langerhans cells,[273] they express class II MHC, CD4 and CD11c. In view of the probable capacity of these cells to present class II restricted antigens to T cells, the interaction between these and other cells is likely to be important in the GC microenvironment. Dendritic cells, probably identical to those described by Grouard et al[272] have been reported as staining for a C-C chemokine (DC-CK1), the production of which is activated by IL-4 and (TNF-α + IL-1).[269] Liu et al[274] propose that the classical $CD14^+CD21^+$ FDCs (expressing DRC-1, KiM4 and 7D6 antigens) are of stromal origin (in agreement with their lack of CD45 expression[275]), while the $CD4^+CD11c^+$ GC dendritic cells represent hematopoietic cells that may be analogous to the antigen-transporting cells described in mice[176,276] and in chickens.[277]

In addition to the production of chemokines, these cells also produce a number of cytokines, similar to the $CD14^+$, macrophage-like, subset of dendritic cells identified recently as differing from the $CD1a^+$, epidermal associated, subset.[278] In situ hybridization studies have identified a desintegrin metalloproteinase, called decysin, which is highly represented within the DC in GCs.[279] This family of proteinases has the ability to cleave TNF, FasL and CD30 from the cell surface[280,281] and might, therefore, play an important role within the GC microenvironment.

T-GC B Cell Interactions

The exact nature of the role of T cells in the induction and/or maintenance of GCs remains to be determined. It is not likely that the T dependence of GC formation is solely at the level of the initial activation of B cells during the primary antibody response, since antigens such as *B. abortus* induce quite a good primary Ab response but no GCs in the absence of T cells. In view of the observation discussed above that the majority of CD4$^+$ T cells located in a specific GC bear the Vβ and Vα chains of TCR specific for the antigen inducing that GC, it is likely that GC T cells which get activated by appropriate antigen presentation locally influence the proliferation, isotype switching, survival and somatic mutation rate in the GC B cells by contact and/or cytokine production. Indeed, GC B cells are very effective antigen presenters to T hybridoma cells in vitro[186] and may be expected to stimulate the antigen-specific T cells in their vicinity in vivo. The importance of CD40-CD40L interaction in the process of GC formation[282] is in agreement with this interpretation. The antigen specific CD4$^+$ T cells in GCs not only exhibit expression of CD40L,[56,126,283,284] they also possess a memory phenotype (CD45RB$^-$)[285] from the first moment GCs are produced during the immune response. In the mouse, these T cells are Thy-1^{lo}, perhaps also as a result of their activated state and/or memory phenotype.[286,287] T cells are also found associated with GCs produced under abnormal conditions in nonlymphoid organs.[288]

As discussed above, the production of cytokines by GC associated T cells may be important for the local, in vivo maintenance of B cell proliferation, but there is no formal proof for a major role of any given cytokine in this respect. IL-4 appears very important for the continued generation of centrocytes from centroblasts in vitro, and continued stimulation via sIg in the presence of IL-4 promotes the generation of memory B cells rather than plasma cell formation from GC cells.[289] In the analysis reported by Kelly et al,[285] it was estimated that only approximately 1% of the GC T cells contained detectable mRNA for IL-4, although a large percentage of these T cells were proliferating, as shown by incorporation of BrdU. It is remarkable that even murine EL-4 lymphoma cells, which in essence behave like activated T cells of Th2 phenotype, are capable of maintaining human GC B cell proliferation. It was estimated that, in cultures of isolated human GC B blasts with EL-4 cells, a net increase in B cells is observed within 4 days, while no viable B cells are found after 24 hours if EL-4 cells are omitted.[290] These investigators also report that the percentage of CD77$^+$ B cells in such cultures remains high over the first 4 days, after which 1/4 of the cells differentiate to Ig secreting cells by day 10.

On the other hand, studies on the effect of anti-CD40L and of anti-B7-2 on GC formation strongly suggest that cell interactions requiring contact via these molecules and their ligands are important for the maintenance of GCs.[207] In some of the experimental designs involving knockout mice or injection of Abs to these interaction molecules before or immediately after antigen,[282,291-294] the interpretation of any effect obtained on GC formation is complicated by the additional need for T-B cell interactions mediated by these molecules in the outer periarteriolar lymphatic sheaths and around terminal arterioles during the early part of the T-dependent immune response.[295] Any interruption of this part of the response might indirectly interfere with GC formation, since B cells may need to be activated outside of GCs before they can participate in the GC blast cell proliferation. In addition, both anti-B7 and anti-CD40L can interfere with the T cell priming[296,297] that may be needed for GC formation. Moreover, injection of soluble CD40-IgG1 fusion protein is not as effective as anti-CD40L in blocking GC formation in vivo, although it greatly decreases Ig isotype switching in the primary response.[298] In other experiments, it was found, however, that anti-CD40L and anti-B7-2 can interrupt and/or abolish GC formation when they are injected as late as 6 days after the primary antigen injection and GC formation is analyzed on day 12.[207] Under those conditions an effect of the Abs on the primary response is avoided. In view of the expression of B-7 on FDCs[55,56] as well as GC B cells, and of CD40L on some activated B[299] as well as activated T cells, it cannot be automatically concluded that the interruption of GC formation in the treated mice is necessarily due to interference with T-B cell interaction. In fact, the low numbers of T cells seen in the dark zones of GCs makes it difficult to propose a need for T-B cell interaction in the maintenance of proliferation in centroblasts. It should also be noted that the disruptive effect of anti-CD40L on GC formation in vivo is not mediated by induction of apoptosis of B cells.[300] However, recent studies suggest that the T cells in GCs regain the sensitivity to apoptosis of immature thymocytes and become apoptotic after injection of anti-CD3, dexamethasone or bacterial superantigen.[301] It is possible that the same is true after injection of anti-CD40L.

Interactions between GC B cells and CD40L$^+$ T cells is probably of great importance for the survival as well as the differentiation of the B cells. In vitro exposure of GC B cells to anti-CD40 increases Bcl-2 expression along with increasing the survival of the cells.[302] Exposure to CD40L causes a >80% survival of GC cells after 1-2 d, as compared to <10% in control cultures. However, Bcl-2 induction comes up too late to account by itself for the protective effect of CD40L.[228] In addition, it should be emphasized that, even in the total absence of *bcl-2* (in *bcl-2$^{-/-}$* mice), CD40L expression on T cells still causes protection of B cells from spontaneous apoptosis in vitro.[229] Of greater significance in this respect is the observation that CD40 engagement on normal murine B cells by CD40L causes a marked increase in the expression of the Bcl-2 homologue, Bcl-X, and not in Bcl-2 itself.[303] It is therefore more likely that the rapid induction of Bcl-X_L by CD40L, and in particular by the simultaneous engagement of sIg and CD40, is of major importance in the protection of tonsillar B cells against apoptosis afforded by these agents.[304] Marked synergy in the upregulation of Bcl-X_L by anti-Ig and anti-CD40 is also seen in studies on small resting B cells[305] and on Burkitt's lymphoma cells.[306] In the studies with resting B cells, synergy between anti-Ig (or anti-CD40) and IL-4 in causing Bcl-X_L upregulation was also seen.[305] In RAMOS cells, IL-4 does not induce Bcl-X_L upregulation, but it inhibits the appearance of the pro-apoptotic protein Berg 36 during apoptosis induced by ionophores.[307] It should also be realized that studies such as those by Billian et al[308] and Galibert et al[309] suggest that, in a subpopulation of GC cells, the sensitivity to apoptosis is enhanced by interaction of sIg with ligand subsequent to stimulation via CD40. Bcl-X levels have not been studied in such a subpopulation, but the effect of sIg triggering in these cells is counteracted by IL-4 (presumably T help!).

Both the CD40-mediated and the sIg-mediated rescue of GC cells from apoptosis require tyrosine phosphorylation,[310,311] whereas, in contrast to the effect of anti-CD40, that of anti-Ig is inhibited by TGF-β.[245] Indeed, it was shown recently[312] that phosphorylation of Thr234 in the cytoplasmic domain of CD40 causes the induction of A20, a novel zinc finger protein which renders B cell lines resistant to apoptosis. This A20 expression is mediated by inducible binding of NFκB complexes to the A20 promoter. One of the mechanisms by which triggering with anti-CD40 and/or other ligands for GC cells may cause resistance to apoptosis is through the induction of IL-10 production in these cells, as it has been shown that the addition of exogenous IL-10 to large splenic B cells in culture causes a 2- to 3-fold higher recovery of viable cells and a strong expression of Bcl-2 protein in the cells.[313]

The susceptibility of centroblasts to apoptosis is associated with low content of cAMP,[311] but induction of an increase in intracellular cAMP with forskolin does not affect the spontaneous apoptosis of GC B cells.[314] Experiments with EBV transformed human B cell lines have shown that transfection with *c-myc* changes the phenotype of the cells into the direction of GC cells.[315] The cells become very sensitive to apoptosis, but can be rescued by exposure to anti-CD40 or CD40L, after which they exhibit enhanced CD10, CD38 and adhesion molecule expression, while CD39 and CD23 go down. These results suggest that c-Myc expression may cause the acquisition of centroblast features. It has been observed that ligation of HLA class II molecules promotes the sensitivity of immature B cells to CD95-mediated apoptosis.[316] In addition, CD95 ligation on GC B cells antagonizes the stimulatory effect of anti-CD40 on cell proliferation, survival and Bcl-2 expression.[317] Conversely, coculture with CD40L transfected cells prevents the apoptosis induced by anti-CD95 in low density tonsillar B cells and in the CD95[+] RAMOS cell line of GC phenotype.[318] However, the situation is more complex than a straight-forward counteraction between these receptors, because interaction with CD40L-expressing T cells causes the up-regulation of Fas, CD44 and CD62L (=ligand for P-selectin), while downregulating CD10, CD38 and CD77 on GC cells; anti-CD40 cannot mimic most of these effects.[317] Upregulation of Fas on RAMOS cells by interaction with CD40L[+] T cells has also been reported.[319] These investigators find that resistance to apoptosis is obtained by simultaneous triggering of the cells with anti-Ig and anti-CD40, although there still is upregulation of Fas. It is of interest that CD77, which is a globotriaosyl ceramide, mediates apoptosis when interacted with its ligand verotoxin[320] and is downregulated by interaction of GC cells with anti-CD40.[321]

Triggering via CD38 is known to result in costimulation with anti-Ig[322,323] and in protection against apoptosis[324,325] (see Table 1.4). Indeed, such signaling activates the tyrosine kinase Syk and phosphorylates both phopholipase C-γ and phosphotidylinositol-3 kinase.[326] A ligand for CD38 has been identified, moon-1, which is expressed on many cell types.[327] It is of interest that ligation of moon-1 on T cells causes an increase in intracellular Ca^{2+},[327] suggesting the possible relevance of the ligand pair moon-1-CD38 for the T-B cell interaction in human GCs. It should be realized, however, that such interactions cannot be of importance in murine GCs, as GC B cells and plasma cells are CD38[−] in mice.[50] On the other hand, in mice,

Table 1.4. Interactions affecting apoptosis of germinal center cells

B Cell Surface Marker	Interacting Molecule on: T Cells	FDC	Effect on Apoptosis (Ref) /Mimicked By
sIg	TCR	Ag/IC	↓[222,244,245]/Anti-Ig
Ia	TCR/CD4	–	?/Causes cAMP rise[592,593]
CD19 (with CD21 or CD77)	?	IC + C	↓[245]/Anti-CD19
CD21 (=CR2, with sIg)	–	IC + C, CD23	↓[594,595]/Anti-CD21; C3 +Ag; CD23*
CD20	?	?	↓[596]/Anti-CD20
CD38	Moon-1**	?	↓[324]/Anti-CD38
CD40	CD40L	–	↓↓[222,228,244]/Anti-CD40; CD40L
TNF-R	sTNF-α or β		↓[597]***
LFA-1	ICAM-1	ICAM-1	↓[225]/Soluble rICAM-1; Anti-LFA-1
VLA-4	–	VCAM-1	↓↓[225]/Soluble rVCAM-1
Additional unidentified		Unidentified, none of the above	↓(Rescue from anti-CD95-induced)[598]
CD77	?	?	↑[320] /Verotoxin
Fas (CD95)	FasL		↑[317,318]/Anti-Fas and Anti-CD40/ CD40L: opposing effects

* CD23 is a ligand for CD21, CD11b and CD11c.[599] However, soluble CD23 was shown to protect human pre-B leukemia cells from apoptosis, even though these cells were CD23⁻, CD11b⁻ and CD11c⁻.[600]

** A ligand for CD38 has been identified, moon-1, which is expressed on many cell types,[327] but it is not clear whether this is the only ligand or the one responsible for the effect of CD38 ligation on apoptosis.

*** It should be noted that these studies were performed with Burkitt's lymphoma cells because these cells are thought to be representative of GC cells. There are, however, some significant differences between the responses of GC cells and Burkitt's lymphoma cells such as RAMOS. While RAMOS cells are induced to undergo apoptosis with anti-IgM,[601] GC cells are protected from apoptosis by anti-Ig (see above). It is true that GC cells could be heterogeneous in their properties, for instance there might be significant differences between subsets such as centrocytes and centroblasts. In view of the lack of information about this point at this time, the data on GC cells have been summarized here as if representative of a single population.

ligation of CD38 could be of importance in the activation of resting B cells towards GC development, as triggering via CD38 induces IL-5R upregulation and, like IL-5, requires the presence of Btk for its effect on B cells.[325] As shown in Table 1.2c, Btk$^{-/-}$ mice have decreased GC formation.

Possible B-B Cell Interaction in GCs

Freshly isolated GC B cells themselves do not express mRNA for any of 10 cytokines examined.[328] Corcione et al[329] found G-CSF production in GC B cells, which was enhanced by stimulation with anti-CD40 + IL-4. IL-10 production, although frequently found in activated blast type B cell lymphomas, is also absent from Burkitt lymphomas exhibiting the GC-like phenotype.[330] In situ staining for cytokines has so far failed to detect IL-4, IFN-γ or IL-2 within follicles 7 days after footpad injection of antigen.[295,331] Further studies on this subject are clearly needed. Expression of CD40L on activated human B cells has been reported[299] and could of course result in B-B cell interaction in the closely packed centroblast areas. However, any CD40L expression found by staining GCs in situ has been in T cells.

Nature of Cytokines Required for Germinal Center Induction

Relatively little is known about the cytokine requirements of GC B cells for proliferation and/or differentiation and the role of GC T cells in this respect. CD77$^+$ low density and/or PNA$^+$ cells from human tonsils proliferate in response to the combination of anti-CD40 and IL-4, but show very low or no response to anti-IgM or *Staphylococcus aureus* Cowan I (SAC) and IL-2 or IL-4. This is in complete contrast to PNA$^-$ B cells which respond much better to anti-IgM or SAC with cytokines than to anti-CD40.[38,332] None of several other cytokines, including IFN-γ, IL-5 or LMW-BCGF, costimulate the PNA$^+$ cells with anti-CD40. Lagresle et al[333] confirmed these findings and found in addition that the typical GC B cells were also not induced to secrete Ig by any of the stimuli, while typical memory phenotype (IgM$^-$IgG$^+$CD38$^-$CD39$^+$CD10$^-$) B cells proliferate vigorously and produce large amounts of IgG in response to SAC or anti-CD40 and cytokines such as IL-2, IL-4, or IL-10. Recently, promoting effects on the proliferation of CD38$^+$CD77$^+$ CD44$^-$ GC B blasts have been reported for IL-7 in synergy with IL-10 and IL-3 or IL-4 and for IL-6 in synergy with IL-3 and IL-10.[334] It thus seems that for human centroblasts IL-10 may be a most important ingredient in these mixtures. IL-10 also enhances the viability of large splenic B cells, mostly representing GC cells, by increasing their resistance to apoptosis.[313] Although IL-13 shares most of the effects of IL-4 on B cells, and like IL-4, synergizes with anti-CD40 in the stimulation of B cells,[335,336] no studies specifically addressing the effect of IL-13 on GC B cells have been reported. As discussed in chapter 5, however, many GC-derived lymphomas express[337] and/or respond to IL-13.[338]

Proliferative responses of murine PNA$^+$ and PNA$^-$ B cells from lymph nodes draining the site of antigen injection have been compared, using thymidine incorporation and colony formation in agar. In both assays, murine rIL-5 and human LMW-BCGF caused a much greater costimulation with dextran sulfate of PNA$^+$ than of PNA$^-$ cells.[339,340] Although peritoneal B cells also responded well to IL-5 plus endotoxin and dextran sulfate, the addition of IFN-γ inhibited them, whereas it enhanced the responses of GC B cells.[341] Chondroitin sulfate C, which weakly stimulates murine B cells in the absence of added cytokines,[342] in the presence of IL-4 stimulates PNA$^+$ but not PNA$^-$ B cells.[340] It is of interest to note that IL-5 may be an important cytokine for GC B cells in mice, although there is no convincing evidence that IL-5 affects human B cell proliferation.[343]

In studies on reconstitution of GC formation in nu/nu mice, it was found that cloned T cells with specificity for the antigen used to induce GCs help GC formation in draining lymph nodes, particularly if they are injected into the same feet as the antigen; when injected i.v. only, they primarily help the induction of GCs in spleen.[344] On comparison of typical Th1 and Th2 clones in their relative abilities to reconstitute GC and antibody formation of various isotypes, an interesting discrepancy is found. While Th2 cells alone are capable of significantly raising the number of GCs above background, the combination of Th2 + Th1 cells is generally more effective, while the effect of Th1 cells alone is not significantly different from background. In contrast, Th1 cells reconstitute antibody formation of the γ2a isotype at least 10-fold higher than Th2 cells and 3-fold higher than the combination of Th1 + Th2 cells. Th2 cells alone cause the appearance of IgG1 and IgE Abs, at least 15-fold higher than Th1 cells alone and 2- to 8-fold higher than Th1 + Th2 cells combined.[344-346] Thus with respect to isotype switching, the two types of clones appear to interfere with each other rather than synergize, as they do in the formation of GCs. It is of interest to note that synergy between Th1 and Th2 clones in the induction of GCs in vivo is in agreement with the observation that IFN-γ and IL-5 synergize in the induction by GC B cells of CFU formation in vitro.[341]

Typical Th1 cytokines that have both autocrine and paracrine growth promoting and differentiating effects on human B cells are TNF-α and LT.[252,347-350] Herpes virus Saimiri transformed human T cell clones express TNF-α which causes polyclonal activation of B cells in an antigen nonspecific, MHC-unrestricted and contact dependent manner. This activation involves cell surface TNF-TNF-R and CD2-CD58, but not CD40-CD40L interactions.[351] Although some of these studies were performed with tonsillar B cells, a GC origin of the responding B cells was not established in these studies.

Studies with cytokine transgenic or knockout mice have so far not resolved the question of which cytokines are needed for GC B cell proliferation in vivo (see Table 1.2b). With respect to Th2 cytokines it has been reported that IL-4$^{-/-}$ mice of C57BL/6 background exhibit few GCs in their gut-associated lymphoid tissue, while other peripheral lymphoid tissues do have GCs,[352] suggesting that IL-4 plays a role which can be replaced by other factors. In IL-4$^{-/-}$ mice of BALB/c background, a decrease in GCs in Peyer's patches is not nearly as obvious (Rizzo L, Simmons WJ, Thorbecke GJ, unpublished observations). In view of the role of various STATs in mediating cytokine signal transduction (reviewed by O'Shea[353]), examination of certain stat$^{-/-}$ mice is also of interest. Recently, we have examined stat4$^{-/-}$ and stat6$^{-/-}$ mice of partial BALB/c background. Both exhibit normal GCs, both in Peyer's patches and in lymph nodes draining the site of antigen injection (Table 1.2c; Grusby MJ, Simmons WJ, Thorbecke GJ, unpublished observations), suggesting that neither IL-12 nor IL-13 and IL-4 are essential for GC formation. IL-5$^{-/-}$ mice also do not appear to have defective GC formation.[354] In contrast, IL-6$^{-/-}$ mice have reduced GC development[354] and GCs are hyperplastic in IL-6 transgenic mice,[355] suggesting the possible importance of this lymphokine. However, addition of IL-6 to Peyer's patch B cells in vitro does not enhance their proliferation, even though it induces differentiation of memory B cells into IgA secreting plasma cells.[221] Since there is no direct evidence that any of the other cytokines have a growth promoting or inhibiting effect on murine GC B cells, further studies on cytokine knockout, including the very interesting LT$^{-/-}$ and TNF$^{-/-}$ mice, are discussed under genetic models later in this chapter.

In view of the prominence of CD57$^+$ T cells in GCs, it is of interest to consider the ability of these cells to produce cytokines. Surprisingly, whether isolated from peripheral blood or from tonsils, these cells are apparently not very responsive, produce little cytokine on stimulation with phytohemagglutinin or pokeweed mitogen[121,356] and fail to promote Ig production in B cells.[129,357] Despite this, NK57+CD4+ T cells appear equally efficient as CD57−CD4+ T cells at rescuing GC B cells from apoptosis in vitro.[358] Moreover, purified human GC T cells from tonsils were found to contain mRNA for IL-4 and occasionally also for IL-10, TNF-α and IFN-γ prior to stimulation, suggesting that they had already been stimulated in vivo and produced cytokines of the Th0 (or both Th1 and Th2) subsets.[328] In addition, the NK1.1+ CD4+ T cells in mice are excellent producers of IL-4[133,359,360] and may play a role in the IL-4 production that causes induction of IgE synthesis in vivo after injection of anti-IgD.[361] These cells have also been implicated in the early burst of IL-4 production in BALB/c mice which promotes Th2 rather than Th1 helper cell development in this strain in response to certain parasites.[362] As discussed above, however, these cells are not needed for GC formation in mice (see also Table 1.2a).

In Vitro Induction of Germinal Centers

Stimulation of B cells via anti-CD40 causes homotypic aggregation that is slow in onset, enhanced by IL-4, and largely abrogated by anti-LFA-1 Ab.[363,364] This finding has been interpreted as suggesting the autocrine production of factors that promote B cell activation. Indeed, it has recently been demonstrated that anti-CD40 induces the production of LTα and TNF-α,[251] cytokines known to stimulate B cell proliferation.[348] Stimulation via CD40 also induces upregulation on the B cells of CD54, B7-1 and B7-2.[365,366] Addition of anti-CD3-stimulated CD4+ T to B cells has been reported to increase the PNA binding capacity of the B cells, suggesting the acquisition of a GC B cell like phenotype. This phenomenon is dependent on CD40-CD40L interaction.[367] However, stimulation with anti-CD40 and IL-4 also causes upregulation on B cells of CD23, a surface marker that is absent from GC cells.[366,368] It appears, nevertheless, that the stimulation via anti-CD40 + cytokines preferentially stimulates B cell proliferation and memory phenotype development rather than differentiation into plasma cells.[369,370]

With human B cells, Bancherau and coworkers[371,372] have been unable to get convincing evidence that a complete GC phenotype may be induced in the cells that includes CD10 and CD77 expression as well as somatic mutation, although anti-CD40 + IL-4 and/or IL-10 induces B cell proliferation and Ig isotype switching. Nevertheless, a partial GC B cell phenotype has been obtained in several studies.

The simultaneous triggering of resting human tonsillar B cells via CD40 and sIg results in activated B cells expressing CD38, CD95, carboxypeptidase-M, CD71, CD80 (B7-1) and CD86 (B7-2), resembling in many respects GC cells, but these cells still do not become CD44⁻ and CD10⁺.[373] In addition, such costimulation of the sIg, CD40 and also CD19 induces some expression of CD77.[374] Like GC cells, these activated B cells lack CD24 (HSA). Similar changes are induced in cord blood B cells by coculture with irradiated, PMA + PHA activated EL-4 murine thymoma cells.[375] The blast cells arising after 5 days in such cultures are IgD⁻CD5⁻CD38⁺CD77⁺CD95(Fas)⁺ and Bcl-2^{lo}. Like GC B cells, they are sensitive to Fas-induced apoptosis. In addition to CD40-CD40L interaction, it has recently also been suggested that CD30-CD30L interaction may be important for induction of T-dependent B cell proliferation and/or differentiation, but its relevance to GC formation has not yet been evaluated.[376]

The methods for isolation of FDCs, pioneered by Szakal and coworkers,[232] and the availability of cell lines derived from FDCs, have made in vitro studies on cellular interactions between FDC and other cells possible. In vitro more or less fibroblastic cell lines with some properties similar to human FDC have been derived from freshly isolated FDC by transformation with Epstein-Barr virus[377] or by hybridization with a murine myeloma line.[243] Low density cells from human tonsils gave rise to a cell line, FDC-1, that expresses CD40, CD54, CD73 and CD74 as well as receptors for nerve growth factor.[378] Reasoning that T-B cell interaction alone may not be sufficient for GC development in vitro, investigators have added isolated ex vivo FDC, or cells from these FDC-like cell lines[379] to the cell mixtures (reviewed in ref: 274). Low density B cells isolated from immune mouse lymph nodes form large aggregates with FDCs that need the presence of T cells to continue proliferation. Indeed, there appears to be a requirement for specialized T cells here, as the memory B cell phenotype is induced by coculture of GC B cells with autologous GC T cells of memory phenotype in a CD40L dependent fashion, but not by CD40L expressing activated T cells of naive phenotype.[380] The FDC-B-T cell cluster formation and proliferation is dependent on the expression of ICAM-1, LFA-1, CD44, class II MHC, and sIg.[381] The FDC-like cell line, HK, also binds tonsillar sIgD⁻CD38⁺ B cells and protects them from apoptosis, as discussed above.[226] Stimulation of proliferation and differentiation of memory B cells via iccosomes occurs in these cell aggregates, but is only effective when FDCs are present in addition to the iccosomes.[382] Quite another question is whether FDCs can facilitate the induction of GC B cell characteristics in resting B cells. It has been observed that resting B cells also adhere to FDCs[378] and that FDCs have a chemotactic effect on B cells.[383] While FDCs enhance the proliferation induced by anti-CD40 + cytokines in high density B cells, typical GC cells are not seen in the cell clusters that are obtained in these cultures.[371,372] However, in the presence of antigen (immune complexes) bearing FDCs, resting B cells with specificity for that antigen become excellent antigen presenters to T cells and express an increase in surface B7-1 and class II MHC.[186,384]

Generation of Diversity in V Regions of the Ig Genes in B Cells

Gene Conversion in the Chicken and the Rabbit

The Chicken

During ontogeny and the first three weeks after hatching in the bursa of Fabricius of the chicken, diversification of the Ig repertoire, subsequent to VDJ and VJ rearrangements of a single functional V gene in each of the IgH and IgL loci, is generated through sequential gene conversion of gene segments of the IgH and L chain loci.[385-387] The germline pools of donor sequences are found in families of V pseudogenes located 5' of the functional V gene of each locus. This process occurs in large follicular structures that superficially resemble the large GCs in Peyer's patches, but on closer inspection each consists of a cortex containing blast cells and a medulla containing smaller mature B cells. The generation of naive B cells takes place in the follicles of this primary lymphoid organ, which is equal in size in germfree and normal chickens[151] and exhibits a characteristic involution on sexual maturation. During the antigen-induced immune response chickens, like mammals, produce GCs in their secondary lymphoid organs, the spleen and gut-associated lymphoid tissue. Moreover, their Abs undergo affinity maturation, as do mammalian Abs.[388] It has been reported[387] that point mutations further diversify the chicken Ig repertoire, and a relationship between these somatic mutations and GCs has also been suggested in this species.[389]

The Rabbit

In the rabbit (reviewed by Knight and Crane[200]), the generation of the primary Ab repertoire is dependent on the development of GCs in their gut-associated lymphoid tissue, primarily in the appendix, which is a very large organ in rabbits. This "bursa-equivalent" differs in several respects from the bursa: (1) It requires exposure to a bacterial flora (antigen or superantigen?) for its development; (2) it develops only a few weeks after birth; (3) the tissue contains large typical GCs. However, the situation is otherwise very similar to that in the bursa in that the majority of naive B cells rearrange only a single VH gene and, in these GCs, diversify their Ig repertoire by gene conversion.[24] As the rabbit matures, the appendix appears to evolve from a primary into a secondary (or secondary + primary) lymphoid organ, in which both gene conversion and somatic mutation occur simultaneously in GCs.[201] The question comes up, of course, of whether gene conversion adds to the generation of diversity in GCs from other mammals. So far, convincing evidence in favor of this point has not been presented for the mouse.[390-392] However, in cattle, gene conversion has been found to contribute to Vλ diversity in Peyer's patch B cells.[393]

Somatic Hypermutation in the Sheep Embryo

As will be seen in the following, somatic hypermutation is used in most mammals for the fine tuning of antibody specificity during the development of the memory B cell repertoire and is in those species confined to GCs. However, a major exception to this rule has been found in the sheep embryo. As described above, the B cell follicles in the gut-associated lymphoid tissue of the sheep embryo are considered as bursa-equivalent, primary lymphoid tissue, both with respect to structure and because of their total independence of antigen stimulation.[157,158] However, unlike what one might expect on the basis of the situation in the chicken bursa, the Ig V region diversification in this tissue, which gives rise to the primary (naive) B cell repertoire, arises through somatic hypermutation.[156,394]

Affinity Maturation of Antigen Specific Memory B Cells

Somatic hypermutation

Affinity maturation of Ab during the immune response[395] is the result of a competition by B cells bearing Ab of different affinities for antigen and is indeed more effective during responses induced by low than by high doses of antigen. The site of this competition appears to be the GC microenvironment and the process by which the neighboring B cells within the same GC, even though derived from very few precursor cells, come to differ in the affinity of the Ab on their surface is through Ig gene V region somatic mutation.[396] The mechanism of somatic mutation is still unknown, but data showing the existence of mutational hot spots in both VH and VL genes, as well as in targeted transgenes, may give clues to the enzymatic mechanisms involved.[397,398] Full hypermutation activity requires the presence of both the κ intron and 3' enhancer regions.[399] Recent studies suggest that this may be due to the existence of a linkage between somatic hypermutation and transcription.[400] This subject is discussed in detail in chapter 4.

The kinetics of GC formation during the primary response to such a well studied antigen as a conjugate of the hapten 2-phenyloxazolone (OX) parallel those of the appearance of somatic mutations in the Vκ regions of anti-OX Ab that are associated with increases in the affinity for the hapten.[401-404] In addition, high affinity associated somatic mutations in the CDR1 of the VH region are present in only a small subpopulation of memory B cells late in the primary response to (4-hydroxy-3-nitro-phenyl) acetyl (NP) conjugates,[405] another determinant which initially induces a restricted Ab repertoire, dominated by λ1 chain-bearing Ab expressing the unmutated VH gene V186.2 (of the VH J558 family) in combination with the D segment DFL16.1[406] (of idiotype Ac38/Ac146[407]). A third Ab response that has provided information about the kinetics of the appearance of somatic mutations is that to phosphorylcholine (PC), which induces predominantly Ab containing the VH1, DFL16.1 and JH1 with mutations accumulating in the CDR2 and CDR3 during the mid to late phase of the primary response.[408]

Although the cumulative evidence suggests that, like memory cell generation, the process of somatic hypermutation occurs in GCs,[409] the proof for this hypothesis was not obtained until the elegant analysis, by PCR, of cDNA prepared from isolated GC cells during the response to OX[403,404] and of DNA from cells picked from individual GCs in frozen spleen sections from mice immunized to NP[9]. The latter study also showed that the Ab-forming foci formed around terminal arterioles during the primary response did not exhibit somatic mutations. Similar methods were used by others to show that cells from the marginal zone in human spleens exhibit mutations in their VH genes, whereas mantle zone B cells do not, suggesting that memory B cells tend to accumulate in the marginal zone.[410] Also, in human tonsils follicular mantle zone B cells are clonally diverse and predominantly express germline V genes, whereas GC B cells are clonal and exhibit VH mutations.[411]

Although a high frequency of B cells which exhibit VH and VL mutations that confer high binding affinity are found proliferating in GCs (see also review by Berek and Ziegner[199]), the onset of the somatic hypermutation mechanism may not exactly coincide with that of memory B cell proliferation within these sites. It could be that GC precursor cells start to undergo somatic mutation on activation in the periarteriolar sheaths (PALS).[104,412] Further expansion of these cells and selection of high affinity somatic mutants would then occur within GCs. There appears to be an influence of the site in which the immune response occurs: Somatic mutations are found earlier in lymph nodes draining a local antigen injection than in spleen after systemic administration of antigen.[413] In the absence of CD40L expression in individuals with the X-linked hyper IgM syndrome, GCs are not formed, but somatic mutations are still seen in the Ig V genes, albeit at lower frequency than in normal individuals.[414] On the basis of this observation, the authors suggest that somatic mutations may be initiated in a CD40L independent pathway before entry of the cells into GCs.

Analysis of the numbers of somatic mutations in the bm1-bm5 subsets described in Table 1.1b has been made by Liu, Banchereau, and coworkers.[48,415] The sIgM$^-$ IgD$^-$ CD38$^+$ cell type (bm3 and bm4) exhibit many mutations, are sensitive to apoptosis and proliferate poorly on stimulation with CD40L $\pm$ cytokines, whereas sIgM$^+$IgD$^+$CD38$^-$ (bm1 and bm2) exhibit few, if any mutations. They have identified a transitional cell type that is IgM$^+$IgD$^+$CD38$^+$, is sensitive to apoptosis (low Bcl-2, high Fas) and exhibits somatic mutations in the CDR1 and CDR2, fewer when the cells are medium sized than when they are large blasts. They suggest that these cells represent "GC founder cells".[415]

A detailed analysis of the somatic mutations in the PCR-amplified DNA from B cells isolated from splenic GCs has been made by Kelsoe and coworkers. The analysis of shared and distinct somatic mutations observed within an individual GC allows the construction of clonal genealogies of the cells, illustrating the increasing genetic complexity due to repeated Ig V region mutations between days 0 and 16 after injection of antigen.[416] In addition, careful analysis of somatic mutations in individual cells from GCs suggests that proliferation may both precede and follow somatic mutation within a GC, as evidenced by the presence of unmutated and identically mutated sister cells.[199,417,418] It has also been shown that carrier priming accelerates the appearance of Ig gene mutations in a hapten-specific immune response, showing that a variation in T cell help has a profound effect on the process.[419] In the response to PC-protein conjugates, GCs are produced which stain with Ab to the T15 idiotype characteristic of the anti-PC response. Nevertheless, the frequency of mutations in the V gene associated with the anti-PC response (VH1) in these GCs is low compared to corresponding findings in the responses to NP and OX.[143] The presence of specific Ab within individual GC B cells can sometimes be clearly distinguished from that of immune complexes on the surface of FDC.[420] However, it should be emphasized that this is an important ingredient in the interpretation of these findings, because GCs produced in response to the carrier protein of PC might stain with anti-T15 as a result of the presence of PC containing immune complexes on the surface of FDCs in these GCs, while the DNA of most of the B cells might not encode Ab with anti-PC specificity.

One would expect that a decreased efficiency in GC production and memory B cell generation might be reflected in a lack of maturation of the Ab response through defective V gene mutation. In order to detect such an effect, however, the acquisition rate of the somatic mutations during the response should be studied, as any difference might otherwise be missed. Even when the response is low, as reflected by fewer GCs and a lower production of B memory cells during the first 2-3 weeks after the primary immunization, the ultimate affinity maturation and somatic mutation frequency that has occurred in the memory cells that are present by 5-6 weeks after priming may be indistinguishable from that in controls, as shown for instance by a recent study in mice with the *xid* mutation.[421]

At the peak of activity during the primary response, the average rate of mutations in the V regions is $\sim 10^{-3}$/bp/division,[401,422] but this rate drops to $<10^{-7}$/bp/division at later intervals after priming.[423,424] An interesting question is whether memory B cells undergo further mutations after repeated challenge with antigen. This was analyzed in the response to PC-conjugates and it was found that the frequency of VH1 mutations during the secondary response depends on the level of CD4 T cells in the mice.[143] In the response to OX, Berek and Milstein[401] found that upon repeated challenge with antigen, memory B cells produce responses of increasing affinity and continue to accumulate mutations. As shown by an elegant combination of in vivo and in vitro approaches, a subset of the sIgM⁻ sIgD⁻ (switched) memory B cells that is HSA^{lo}, as also seen for GC precursors before priming, regains the ability to accumulate mutations after challenge, probably by producing GCs again.[425] No further mutations are accumulated, however, during the generation of Ab forming cells from memory cells in vivo or in vitro.[423,425] These findings also confirm the previous observation of Shan et al[426] that Ig switching does not terminate somatic mutation. Once somatic mutation is active in a B cell line in vitro, it can continue in the total absence of T cells and T cell factors, as shown by the hypermutation rate of 0.7 x 10^{-6}/bp/division exhibited by a human follicular lymphoma cell line.[427] However, the addition of IL-2 + IL-4 increases the mutation rate 10-fold in these cells. An even higher somatic mutation rate has recently been observed in a Burkitt's lymphoma cell line, BL2, but in these cells the somatic mutations have to be induced by exposure to crosslinking anti-IgM and activated T cells.[427a] Ex vivo isolated GC B blast-containing spleen cells continue the mutational process only if the cells are restimulated via surface Ig and are cocultured with T cells.[428]

Two recent observations point to a possible role of the microenvironment of GCs in the induction of somatic hypermutation. The first is the observation by Zheng et al[113] that T cells with $V\alpha 11 V\beta 3$ TCR, situated within GCs during the immune response to cytochrome C, exhibit somatic mutations in their $V\alpha 11$ gene, even though TCR V regions do not show such mutations either in the thymus or in peripheral lymphoid tissue. The second example is the extremely high rate of somatic mutations (0.14-1.6 x 10^{-2}/bp) observed in the *bcl-6* genes of GC-derived lymphomas, even though *bcl-6* has no obvious homology to Ig.[79] It has not yet been reported whether *bcl-6* also undergoes somatic mutation in cells other than GC-derived lymphomas.

All vertebrates with jaws have B cells and the majority rearrange their Ig genes somatically.[429] GC formation is restricted to homoiothermic animals.[430] Associated with the ability to produce GCs in their lymphoid tissues, birds and mammals produce Abs of increased affinity in the secondary as compared to the primary response. Nahm et al[431] made the interesting proposal that the characteristically fast rise in serum Ab following a secondary exposure to antigen may provide a survival advantage in warm blooded animals in the defense against pathogens that multiply more rapidly at the higher body temperature. Poikilothermic animals, including amphibia,[429] produce no GCs, perhaps because their B cell system is more like the B1 cells (the $CD5^+$ or peritoneal B cell subset in the mouse) which rarely or never produce GCs,[29] or else because they are missing an essential component such as the FDC. Wilson et al[432] have shown that somatic mutations do occur in Ig genes of *Xenopus*, but the mutants are poorly selected, as shown by the nature of the mutations observed, and the low degree of affinity maturation in the anti-dinitrophenol Ab response.

Positive and Negative Selection of Memory B Cells in Germinal Centers

Positive Selection

In any microenvironment where B and T cells undergo gene rearrangements, somatic hypermutation and/or gene conversion, a high rate of spontaneous apoptosis occurs in these cells. This is for instance true for thymocytes in all species, for the ileal Peyer's patch follicular B cells in sheep[433] and for bursa cells in chickens.[434] The B cells in GCs proliferate extremely rapidly and this high rate of proliferation is accompanied by a high incidence of apoptosis.[38,222] It is thought that the microenvironment within GCs allows for a loss of autoreactive and/or ineffective B cells through apoptosis and for the simultaneous selective expansion of memory B cells whose sIg has a high affinity for the inducing antigen (reviewed in MacLennan et al[17]) recapitulating to some extent for the memory B cell what occurs for virgin B cells in the bone marrow.[435,436] Once memory B cells emerge from the GCs, they can go into the circulation as resting B cells, bearing sIg which is frequently of switched isotype and exhibiting somatic mutations in the V regions of their Ig genes.

It is generally accepted that the high rate of cell death is to a large extent due to the inability of cells that have errors in V gene arrangement and/or sequence to be positively selected. The low Bcl-2 content of GC blasts is presumably involved in their high susceptibility to apoptosis. It is, therefore, of interest that in mice transgenic for *bcl-2* under influence of the Eμ enhancer, and expressing Bcl-2 in GC cells, the VH mutation frequency is reduced by 50% as compared to controls, although effective selection of high affinity mutants does occur. There appears to be a reduced tendency to exhibit multiply mutated sequences, suggesting that nonmutated cells may be retained longer in *bcl-2* transgenics than in control mice.[437,438]

In addition to a gain in affinity, a loss of specificity for the GC-inducing antigen may also be expected after somatic mutation of V genes. Analysis of PNA[+] B cells has failed to identify many such loss mutations, not even when the cells are examined early in the response.[412,413] It is possible that those cells that fail to find a ligand are lost through apoptosis and are rapidly phagocytized by tingible body macrophages. However, it has recently become clear that a process called "receptor editing" allows developing precursor B cells in the bone marrow to use sequential VH or VL chain locus rearrangements to recover from nonproductive rearrangements or to replace what would otherwise be strongly autoreactive rearrangements. According to recent findings, the reexpression in GC B cells of RAG-1 and RAG-2[71,73] allows them to "edit" their receptors[439,440] and thus might salvage B cells in GCs that have destroyed one of their rearranged V regions, in particular their V_L, through unsuccessful somatic hypermutation. On the other hand, a strong signal via sIg, indicating high affinity interaction with antigen, may cause the cells to stop the somatic mutation process and thus constitute part of the mechanism of "positive selection".[199]

Negative Selection

The rescue of cells with high affinity for the antigen localized within the GC suggests a mechanism of positive selection, the default pathway being apoptosis. If the high sensitivity of GC cells to apoptosis would also be aimed at the prevention of autoimmunity caused by cells with mutated sIg binding to autoantigens, one would expect negative selection to occur as well. Indeed, Klinman, Linton and coworkers[425,441-443] have shown that the newly emerging antihapten memory B cells arising after T cell dependent antigen stimulation are susceptible to tolerization in vitro by hapten presented on a carrier not recognized by the primed T cells. It might be argued that negative selection would not be needed, except perhaps at the very early centroblast stage and then only for cells that express sIg, which may not be the case for the majority of centroblasts. At later stages, as centrocytes, the only cells with a chance to attain interaction status with first FDCs and then T cells are most likely those cells that can bind the specific antigen inducing the GC. Those cells will be positively selected, whereas cells of the wrong specificity presumably will not. Thus, one could postulate that there are two phases of selection, first at the centroblast stage in the dark zone, where the cells undergo negative selection if autoreactive, followed by positive selection at the centrocyte stage where the cells are near the (FDCs + antigen) and the T cells, both of which cooperate in the rescue of B cells from apoptosis and the induction of further maturation.

There is no doubt that negative selection and receptor editing occur in primary B cells, as evidenced for instance by observations on mice carrying a rearranged Ig transgene with autoreactivity.[444-446] In mice expressing double transgenes, one a rearranged Ig encoding the H chain of a hen egg lysozyme (HEL)-specific Ab and the other HEL, as a membrane bound antigen on autologous cells, both low and high affinity lysozyme binding B cells are deleted.[447] When HEL-specific B cells from a transgenic donor mouse are injected into a mouse with an ongoing immune response to a cross-reacting lysozyme, administration of soluble HEL at the peak of GC production rapidly eliminates HEL specific GC B cells by apoptosis. When the cells also express Bcl-2 as a transgene, the elimination of these cells is inhibited.[448] Similarly, soluble deaggregated antigen injected during the GC response to that antigen in normal mice causes increased apoptosis in GC cells,[300,449] predominantly in light zone centrocytes and peaking within 4-8 hours after injection. Repeated injections of antigen at the peak of GC formation results in a significant decrease in size and numbers of GCs.[300] The effect is directly on B cells, not primarily mediated by inhibition of T-B cell interaction, and partially prevented by the presence of a *bcl-2* transgene.[449] Apoptosis of GC cells by a high dose of soluble antigen is also induced in *lpr* mice, suggesting that Fas is not essential for its induction,[300] although Nossal,[437] when using a very low hapten conjugate as the tolerizing antigen, does see a reduced effect in *lpr* mice. He suggests on this basis that the expression by the activated B cells in GCs of both Fas and FASL[450] may play a role in this phenomenon.

Anti-HEL Ig transgene-bearing B cells, transferred to a host which expresses HEL as a transgene, are rapidly cleared from the blood and spleen of recipients and are detectable in outer margins of the PALS for ± 24 hours after transfer, but not in follicles. If simultaneous T cell help is provided, the B cells proceed into follicles and produce GCs, suggesting that the availability of T cell help is an extremely important decisive factor in determining whether the B cell will be tolerized and disposed of or activated.[6] Similar observations were made with rheumatoid factor transgenic B cells.[451] Within GCs, a subset of the GC B cells is apparently sensitive to deletion via direct interaction of its surface Ig with antigen or anti-Ig, particularly after having interacted with CD40L and in a Fas independent fashion.[308,309] This effect is counteracted by IL-4 and may therefore also be negated by T cell help. Thus, the effect on GC B cells by very large doses of soluble antigen could be: (1) directly on the B cells, by interaction with sIg without allowing sufficient interaction with neighboring cells, and (2) indirectly, by inactivating the antigen-specific T cells in the follicle and thus actively withdrawing from the B cells the previously enjoyed T cell help. One might ask what helps the B cell distinguish between an autoantigen in response to which it should die, and the GC-inducing antigen in response to which it should mature? Presumably, a competition for its sIg between self-antigen and the GC-inducing antigen will occur which, if won by the self-antigen, will result in a lack of help from T cells, since the T cells are presumably tolerant to the self-antigen. This could, therefore, still be "positive selection" (i.e., death by neglect), but it could also be an active negative selection as a result of killing of the B cell via sIg crosslinking in the absence, but not in the presence, of T cell help (such as protective IL-4, see above).

Morphologically, the apoptosis observed in GCs after injection of large doses of soluble antigen resembles that induced by injection of dexamethasone.[205] Moreover, the dexamethasone induced involution of GCs is also reduced in *bcl-2* transgenic mice.[452] The i.v. injection of antigen into an immunized mouse already producing Ab could represent a stressful event leading to the endogenous release of glucocorticosteroids. It would therefore be of interest to see these results repeated in adrenalectomized mice or else in the presence of corticosteroid receptor blocking agents.

While these experiments suggest that negative selection occurs in GCs and is at least partially prevented by enhanced expression of Bcl-2, it has been found that positive selection and affinity maturation are normal in *bcl-2* transgenic mice which express Bcl-2 protein in their GC cells.[438] Such mice produce a larger incidence of GCs in the follicles of lymph nodes draining the site of antigen injection than littermate controls,[452] but that is perhaps partially the result of their greatly increased peripheral B cell numbers.[453] It is of interest that, although Fas is highly expressed in GC cells in normal mice, *lpr* mice have a normally regulated response to NP conjugates.[67]

Conditions in Which GCs Malfunction

Effect of Aging on Germinal Center Formation and Function
GC formation[454-456] and secondary Ab responses[457] are defective in 2-year old mice of several inbred mouse strains examined, although in some long-lived F1 hybrids (LAF1) this defect is not yet

obvious at that age.[458] In the GCs that are formed in aged mice, somatic mutations during the response to NP are reported to be lacking, and affinity maturation of the immune response is limited to the Ab affinities of germline Ig.[459] Defects in each of the cell types involved in GC formation, primed T, precursor B and FDC, might contribute to this age-related functional decline. Linton[460] has examined the status of B cell precursors of GCs from aged mice and finds them to be, if anything, slightly increased in number, as long as they receive help from primed T cells from young mice. In contrast, antigen retention on FDC and iccosome formation are deficient in aged mice.[454,461,462] Reconstitution of this deficiency by bone marrow transplantation from young into aged mice results in excellent GC formation.[463] Thus, a relative absence of antigen localization on FDCs possibly represents a major contributing factor to defective GC and memory B cell formation in the aged. A more detailed discussion of the age-related changes in FDCs may be found in chapter 2.

With respect to T cell help, GC formation appears to be less demanding than is the T cell dependent primary Ab response.[464] Indeed, in nu/nu mice, reconstitution of GC formation by T cells is obtained without detectable reconstitution of T cells in paracortical areas in lymph nodes.[108] It is thus possible that, in spite of the known defects in T helper function in aged mice, these defects may not have a major impact on GC formation.

Analysis of the effect of age on somatic hypermutation in the Peyer's patches of mice has been performed by Gonzales-Fernandez et al.[465] A plateau in the incidence of mutations is seen when the mice are 5 months old, whereas the proportion of GC B cells in Peyer's patches peaks at 2 months, decreasing thereafter. The authors interpret this finding as suggesting that a successive accumulation of mutations occurs in Ig genes of memory B cells. A decline in the rate of somatic mutation in GCs from aged mice, such as observed by Miller and Kelsoe,[459] might explain the observation of Smith et al[466] that tingible body macrophages are absent from the GCs in aged mice. Tingible body macrophages are thought to derive their appearance from the fact that they are removing the cellular debris from apoptotic B cells in the GC. It might be expected that less mutations would also lead to less apoptosis and thus to less tingible body macrophages. However, a complete absence in the aged mice is still surprising.

Retroviral Infections (AIDS)

Although the presence of viral particles in GCs may contribute to the maintenance of B cell memory, it can also have a destructive influence on the microenvironment of the GC.[467] The cells within the GC do not necessarily have to be infected themselves, since the presence of viral antigens and the immune response to these antigens may cause bystander destruction. In the specialized case of HIV or SIV infection, the GC appears to represent a good hiding place for virus even during stages of the disease when little or no virus is detectable in the peripheral blood.[182] Thus, any $CD4^+$ T cell that comes in this microenvironment may become infected.[468] In addition, whereas normally $CD8^+$ T cells do not localize in GCs, they are attracted to these sites of virus localization,[469,470] presumably because activated and infected B cells (and possibly also FDCs) are presenting viral antigens. The ultimate result of this process is the progressive destruction of FDCs, either by a direct cytotoxic effect of the $CD8^+$ T cells or through a bystander effect.[471-473] These CD8 T cells are strongly positive for granzyme B, suggesting that they are competent cytotoxic cells.[474] Another reason for depletion of GCs in HIV infection is the eventual lack of $CD4^+$ T helper cells which are required for their continued function and development. A more detailed discussion of the effect of retrovirus on FDCs can be found in chapter 2.

Genetic Models, Transgenes and Gene Deletions

$TCR^{-/-}$

Recently, TCR gene "knockout" mice have been studied for the presence of GCs, including $TCR\alpha^{-/-}$ and $TCR\beta^{-/-}$ mice (see Table 1.2a). Mice lacking the TCR β chain expression are incapable of producing GCs under normal conditions of immunization[146,475] (also Tsuji M, Thorbecke GJ, unpublished observations), suggesting that γδ T cells are relatively ineffective in supporting GC development. However, repeated parasite infections can induce rare GCs in $TCR\beta^{-/-}$ mice.[146] In contrast, $TCR\alpha^{-/-}$ mice do exhibit a variable number of GCs in their spleens and Peyer's patches.[476] Even when $TCR\alpha^{-/-}$ mice are raised under germfree conditions, they sometimes exhibit GCs in their spleen, although not in their Peyer's patches.[476] In our own laboratory, we have observed that a few of these $TCR\alpha^{-/-}$ mice have abnormally high numbers of GCs in their mesenteric lymph nodes (up to 60 per lymph node section), but no clear correlation could be found between the degree of colitis development that is seen in these mice and the GC hyperplasia present in their mesenteric lymph nodes (Lakhman Y, Nagler-Anderson C, Thorbecke GJ, unpublished observations). Nevertheless, when attempts are made to induce GCs in such mice by footpad immunization, no GCs are observed in the draining lymph nodes. These mice have, in addition to the TCRγδ cells, some TCRββ or βα (precursor) T cells bearing a variety of Vβs, which are found primarily associated with their GCs.[476] It seems possible that this GC formation is driven by T cells responding to bacterial or viral SAgs or some form of nonprotein antigen.

Eα Transgene Targeted to GCs

In preparing various MHC class II (Eα) transgenic mice, Mathis and coworkers[477] obtained one line in which I-E expression is under control of its own promoter but with a deletion in the 5' flanking region, and, among B cells, limited to GC blasts. This property may be exploited in the isolation of GC B cells, but the cytokines and/or cell interactions needed for the upregulation of this GC specific expression have not been identified, as induction of I-E expression on B cells from these mice could not be obtained in vitro. A region between -1322 and -1180 in the 5' untranslated region of the Eα gene appears to exert a negative influence on this expression, as in its presence and in the absence of the promoter which controls B cell specific expression, Eα is not expressed on GC cells. However, if both the B cell specific promoter and this specific region, which contains an interesting target site termed "W" by the authors, are both absent, GC B cells express I-E.[477]

Mice Lacking Complement Factors or Their Receptors

C3, C4, Cr2 (CD21)

A lack of C3 or C4 results in deficient splenic GC formation in response to i.v. immunization.[478] This defect can be overcome by a 10-fold increase in the antigen dose. T cell priming is normal in such mice, suggesting that the defect is in the B cells or FDCs or both. The effect of C3/C4 absence is reminiscent of older studies in which the antigen localization on FDCs in splenic GCs was disrupted by the transient depletion of C3 with cobra venom,[479] along with a defect in secondary antibody responses.[480] More recently, it was also shown that treatment with anti-C3 prevents the localization of immune complexes on FDCs in the rat spleen and, with it, GC formation.[481] It has not been reported whether the C3 and/or C4 deficiencies also pre-vent GC formation in Peyer's patches and lymph nodes. Cr2$^{-/-}$ mice have a similar defect in humoral immune responsiveness and GC formation to that of C3/C4$^{-/-}$ mice.[482] B cell follicles and B cell re-sponses to stimulation via CD40 are normal in these mice. Studies on chimeric mice in which the FDCs are CD21^{+}, but the B cells are not, or vice versa, have shown that expression of CD21 on the B cell was important for GC formation and for the enhancing effect of complement on the immune response.[483] The effect of CD21 on GCs was suggested to be due to an effect on the survival of B cells within GCs.[484] CD21 may be needed as a costimulator in the CD21/CD19/CD81 complex or else as a ligand for C3d on the immune complexes to establish firm contact with the FDCs. The marked decrease in GCs in CD19$^{-/-}$ mice may be due to a similar defect in the costimulation complex. It should be noted that CD81$^{-/-}$ mice share with CD19$^{-/-}$ and CD21$^{-/-}$ mice a defect in the IgG1 response after immunization with a T dependent antigen,[485] but GCs were not reported on. Not mentioned in Table 1.2b is that the absence of factor B (Bf), a factor required for the initiation and propagation of the alternate comple-ment pathway, has no detectable influence on either Ab or GC formation.[486]

Mice Lacking Cytokines or Their Receptors

TNF-α, LTα and β

Mice lacking lymphotoxin-α (LTα) fail to develop lymph nodes, Peyer's patches or normal white pulp structures in the spleen, in-cluding GCs.[487,488] Mice lacking LTβ, in addition, lack peripheral, but not mesenteric lymph nodes. Lymphoid cells from normal donors do not reconstitute lymph node formation in these mice, whereas LTα or B$^{-/-}$ lymphocytes segregate normally to B and T cell areas in SCID recipients.[489,489a] Thus, although it is not clear what the mecha-nism of this interesting phenomenon is, it seems unnecessary for the lymphocytes to express the LTα gene for their ability to home into already existing lymph nodes. In addition, expression of LTα as a transgene under control of the rat insulin promoter (RIPLT.LT$^{-/-}$), which causes expression of LTα in pancreas, kidney and skin, does restore lymph node formation in LTα$^{-/-}$, but not in LTβ$^{-/-}$, mice.[49c] Additional effects of LT or TNF deficiency on B lymphocyte homing can be detected, as LTα or β$^{-/-}$, TNF-RI$^{-/-}$ and TNF-α$^{-/-}$ mice all exhibit a disorganization in their peripheral lymphoid tissue, most readily noted in spleen, that is characterized by the so-called "ringing of the PALS",[491] meaning a localization of B lymphocytes as a ring around the T cell zone in the periarteriolar region in the spleen, with a variable degree of mixing of T and B cells, resulting in a total absence of B cell follicles.[488,492-496] In the lymph nodes that do develop in $LT\beta^{-/-}$ mice, B cells are located in the cortical area of the node, but in smaller numbers and without follicular accumulations.[497-498] The integrity of the splenic marginal zone structure, i.e., the presence of metallophilic and typical marginal zone macrophages, is much more disturbed in LT than in TNF deficient mice (summarized in Mackay et al[499]). GCs are not seen in the spleens of $LT\alpha^{-/-}$ or $TNF\text{-}RI^{-/-}$ mice, but in lymph nodes from those mice in which some (primarily mesenteric) lymph nodes develop, such as in $LT\beta^{-/-}$ and in rare $LT\alpha^{-/-}$ mice as well as in $TNF\text{-}RI^{-/-}$ mice, GCs are produced. These GCs are, nevertheless, devoid of FDC clusters, as determined by staining with anti-CR1(CD35) or with Ab to FDC-M1.[494,498,500] Similarly, the mesenteric lymph nodes from $RIPLT.LT\alpha^{-/-}$ exhibit GC-like, PNA^+ blast cell areas which lack FDCs.[490] However, on staining for another FDC-specific marker, FDC-M2, a few cells of dendritic morphology are seen in the small PNA^+ areas seen in the spleens of $LT\beta^{-/-}$ mice.[498] Moreover, although the splenic architecture and Peyer's patches are not restored to normal in the $RIPLT.LT\alpha^{-/-}$ mice, their spleens have GCs with FDCs.[490] It is also noteworthy that spleen cells from $LT\alpha^{-/-}$ mice allow production of PNA^+ cell clusters and frank GCs in lymph nodes, but not in spleens, of lethally irradiated $RAG\text{-}1^{-/-}$ recipients.[500] The differences between defects observed in the various lymph node stations, Peyer's patches and spleen in these knockout mice are at this time poorly understood.

The enormous decrease or absence of FDCs in these LT or TNF deficient mice[494,500-502] suggests a necessary influence of both these cytokines on FDC induction and/or local development in follicles. When bone marrow cells from normal donors were used to reconstitute $LT\alpha^{-/-}$ mice, the localization of B and T cells in the recipient spleens was only partially normalized, but GC formation was restored. This correlated with the reappearance of FDC clusters and restoration of IgG responses to a T dependent antigen.[502] Similarly, B cell follicle and GC formation are restored in SCID mice by $LT\alpha^{+/+}$, but not by $LT\alpha^{-/-}$ B cells, regardless of the donor origin of the

T cells.[489a] In the interpretation of results obtained with such reconstituted mice, the FDCs are considered to be predominantly of
recipient origin, which is in agreement with the results of Kapasi et
al.[236] The findings to date suggest that the absence of these cytokines,
LT and TNF-α, results in deficient GC formation as a consequence
of an FDC defect. This could be a particularly prominent feature in
the spleen, where GC formation might be more dependent on FDCs
and the immune complexes fixed on their surface than in mesenteric lymph nodes, where bacterial antigens and/or superantigens
might be present in greater quantities and somehow (with the help
of T cell responses) circumvent the need for antigen localization on
FDCs in GC formation. This reasoning, however, does not explain
the deficient GC formation in Peyer's patches seen in all these LT
and TNF deficient mice (see Table 1.2b).

Further differentiation between developmental and other factors in the effects of LT and TNF-α deficiencies was obtained by
Mackay et al[499] and Rennert et al,[501] who treated mice of various
ages with soluble Ig conjugates of LTβR and TNF-RI (p55). Disruption of most of the lymph node development was found only when
the mice were treated as embryos, although mesenteric and some
other lymph nodes still developed, resembling the LTβ[-/-] mice. Like
mice which possess an LTβR-Ig transgene that is expressed starting
a few days after birth,[503] mice receiving prolonged treatment with
LTβR-Ig as adults had normal lymph nodes, but exhibited the same
disturbed splenic PALS structure as did LT[-/-] mice. In contrast,
whereas prolonged treatment with TNF-R55-Ig produced such an
effect on the splenic architecture in the embryos, it did not do so in
adult mice, suggesting that inhibition of the TNF pathway had only
a developmental effect. Similarly, GC formation was inhibited in adult
mice treated with LTβR-Ig, but not in adult mice treated with TNF-
R55-Ig.[499] It is interesting to note that in LTα transgenic mice in which
LT expression is targeted to a specific organ, ectopic lymphoid tissue containing GCs develops in that organ.[504] Although the observation that GCs are present in lymph nodes from LTα[-/-], LTβ[-/-] and
TNF-RI[-/-] mice shows that none of these cytokines are essential for
GC cell growth, the observations that these factors serve as growth
promoters for activated B cells and are induced in B cells by CD40
triggering, as discussed above, strongly suggest multiple ways in
which they could play a role in GC formation.

Mice Lacking Chemokines or Their Receptors

BRL1/CXCR5

Defective expression of the chemokine receptor, BLR1/CXCR5, causes the absence of lymph nodes and Peyer's patches.[270] This receptor is normally present on mature B lymphocytes and on subpopulations of T cells.[505] It was shown in cell transfer experiments that BLR1$^{-/-}$ B cells failed to enter into follicles of normal recipients' spleens and Peyer's patches.[270] The ligand for this receptor has recently been cloned and named B cell-attracting chemokine 1 (BCA-1).[505a] BLR-1 was first cloned from Burkitt's lymphoma cells.[506] It is closely related to BLR-2 (EBI1),[507] the receptor for a recently cloned member of a novel lymphocyte specific chemokine family located on chromosome 9p13.[508,509] A possibly related defect of B cells to mature into recirculating follicular B cells, associated with defective BLR1/CXCR5 expression, is found in syk$^{-/-}$ mice.[509a]

The *aly* mutation is an autosomal recessive mutation in C57BL/6 mice, called alymphoplasia,[510] characterized by the absence of lymph nodes and Peyer's patches. The *aly* locus maps to a different chromosome (#11) from the TNF/LT locus (#17), suggesting that this is a different defect. However, it could still be related to BLR1 and its ligand. The immune response in these mice is weak, isotype switching is defective, and both GC formation and somatic hypermutation in the response to hapten-protein conjugates (NP-HGG) are absent.[511]

Mice Lacking Transcription Factors

NFκB2/p52, Bcl-3 and Family

In view of the fact that triggering of both TNF-RI and LTβR activate NFκB, it is of interest that the phenotype of mice lacking certain members of the NFκB family, NFκB2 (p52)$^{-/-}$,[512] and Bcl-3$^{-/-}$ mice[513] strongly resembles that of the LTβR$^{-/-}$ and TNF-RI$^{-/-}$ mice (Table 1.2c). They have very poorly developed B cell follicles, FDCs, and GCs, and the morphology of the spleen shows the same "ringing of the PALS" and abnormalities in marginal zones, including the absence of metallophilic and marginal zone macrophages. As yet, only one of the family members, c-Rel, was described as being highly expressed in GCs and marginal zones,[514] and its absence was not reported as having the same effect on lymphoid tissue morphology, although it is not clear to what extent GCs were examined.[515,516]

Absence of NFκB1 (p50) causes a much milder defect in B cell prolif-eration than does absence of p52, while a lack of RelB causes inflam-matory changes with predominantly T cell infiltrates.[517,518] Double knockout phenotypes are now also being studied[512,519,520] and may provide further insight into the cell specificity of the role of some of these transcription factors. B cells from knockout mice for mem-bers of this family that cause lethality either embryonally (RelA[521,522]), or very early after birth (IκB)[523] have so far not been studied for their ability to produce GCs in chimeras.

Bcl-6

One of the most fascinating transcription factors involved in GC formation is Bcl-6. Although mRNA for this protein is present in resting B cells, protein expression in resting cells is low and plasma cells lack Bcl-6 altogether.[80] In contrast, GC cells, both centroblasts and centrocytes, exhibit high levels of this protein, much higher than any other cell in lymphoid tissue.[76,524,525] This protein appears to play a specific role in GC formation, because Bcl-6[−/−] mice completely lack GCs due to an intrinsic lymphoid cell defect,[524-526] while they have an otherwise normal appearance of their lymphoid tissue, in-cluding normal B cell follicles. Cells expressing a truncated form of the Bcl-6 are present in follicles in the spleen, but fail to develop the PNA-binding phenotype or to proliferate in response to immuniza-tion.[525] Responses of B and T cells to various stimuli are also normal in these mice, with the exception that a higher Th2 cell response with eosinophilia is observed in these mice.[526] As might be expected, the lack of GCs in Bcl-6[−/−] mice is accompanied by a complete lack o affinity maturation in the immune response to NP-KLH.[525] It wa: shown by Dent et al[526] that Bcl-6 binds to a consensus binding sit for Stat6 and can, in an in vitro reporter gene system, block IL-4 induced stimulation of transcription. It may therefore be postulate that Bcl-6 is somehow needed in the decision process of GC cells to either continue proliferation (in GCs) or to differentiate (to memor B cell or plasma cell), by modulating the response to IL-4 and/o other stimuli.

Other

Many of the conditions associated with decreased or absent G formation also feature decreased B and T cells, both in number an in function. These include Bcl-2[−/−] and LSIRF/IRF4[−/−]. It is not clea in such cases, without further study, which of these deficiencies contributes the most to the defect in GC formation. It is striking, however, that OCA-B$^{-/-}$, IRF4$^{-/-}$, Lyn$^{-/-}$ and Btk$^{-/-}$CD40$^{-/-}$ mice all feature severe decreases in mature phenotype, long-lived B cells with or without increases in B1 cells (see Table 1.2c). Since CD5$^+$ (B1) and CD24^{hi} B cells are much less capable than mature (CD24^{lo}IgD^{hi}) B cells at producing GCs in SCID recipients,[29] this decrease in mature B cells and the inability of Lyn$^{-/-}$ B cells to respond to triggering via CD40[527,528] may be responsible for the decreased GC formation. In this respect, it would be interesting to know whether NFATp$^{-/-}$ mice exhibit GCs in their lymphoid tissue, since their T cells are defective in CD40L expression after stimulation, while their B cells do respond to anti-CD40.[529] However, so far no report on GC formation in these mice has been published. With respect to Lyn$^{-/-}$ and Btk$^{-/-}$ mice, it is also important to mention here that triggering via both CD38 and IL-5R is LYN and Btk dependent, LYN functioning upstream of Btk activation.[530]

Mice Lacking Cell Surface Proteins

CD40, CD28, B7-1, B7-2

In spite of the fact that triggering of B cells via CD40 upregulates TNF and LT production,[531] the absence of GCs in CD40 mice shows a very different picture from that of LT$^{-/-}$ mice. In CD40$^{-/-}$ mice FDCs are not lacking and the splenic architecture is not disrupted. It may be relevant that the CD40 mediated effect on B cells is tyrosine kinase, but not protein kinase C dependent[532] and involves jak 3 and stat 3, but perhaps not NFκB activation.[533] Triggering via CD40 does seem to activate a special program of gene expression in B cells, as shown also by the accumulation of relB in the nucleus of CD40-, but not of LPS- or anti-Ig-stimulated B cells.[534]

The absence of GCs in peripheral lymphoid tissue in response to immunization, as well as the lack of GC formation in Peyer's patches in CD28$^{-/-}$ mice, suggests that T-B cell interaction leading to activated T and B cells is absolutely required for GC formation and that CD28$^-$ T cells are incapable of activating the B cells sufficiently. Indeed, it is possible that CD28$^{-/-}$, CD40L$^{-/-}$, CD40$^{-/-}$, and Ia$^{-/-}$ mice each fail to produce GCs for comparable reasons (Table 1.2a). The absence of both B7-1 and B7-2 also leads to a complete absence of GCs in the immune response, whereas absence of B7-2 alone prevents

GC formation in response to antigen without, but not with complete Freund's adjuvants (which induces B7-1). Absence of B7-1 alone has no effect.[535]

CD24

CD24[-/-] mice have not yet been reported. Interesting observations have been made on mice transgenic for CD24 under control of the TCR Vβ promoter and the Igμ enhancer,[536] suggesting that this surface molecule is very important in regulating responsiveness of both T and B progenitors to stromal factors such as IL-7 in bone marrow and thymus. In view of the higher propensity of CD24[lo] than of CD24[hi] B cells to produce GCs,[29] it will be interesting to determine the ability of these mice to produce GCs.

Drug Effects

Inhibitory Effects

As already mentioned above, GCs are quite sensitive to antimitotic drugs, such as cyclophosphamide[537] and colchicine,[538] whole body γ-irradiation[205] and prednisolone.[538,539] All of these agents have a strong inhibitory effect on GC formation and memory cell generation, even when given 1 week after the primary antigen injection, i.e., under conditions where the primary response is minimally affected.[538]

Stimulatory Effects

Few drugs have been described that cause hyperplasia of GCs. AGM-1470 is a synthetic analog of fumagillin, derived from *Aspergillus fumigatus,* that has strong inhibitory effects on angiogenesis.[540] D-penicillamine is a 4-carbon fragment of penicillin-G that is used to lower lead, mercury, gold and copper in body fluids and that has as one of its side effects the induction of autoimmune disease.[541] Both of these drugs induce a preferential increase in B cells and GC proliferation, unilaterally in lymph nodes draining the site of injection, suggesting that they induce a poorly regulated, local immune response.[540,542]

Conclusions

GCs are clusters of antigen-induced rapidly proliferating B cell blasts located in primary follicles of peripheral lymphoid tissue. Their peak development is at 7-10 days after antigen exposure. The generation of memory B cells occurs in GCs, accompanied by frequent Ig isotype switching, Ig gene V region somatic hypermutation and/or, in rabbits, gene conversion. The cell types that make up the microenvironment of the GC, their interactions and cytokine production are of great interest. The cell types include two types of B cells, centroblasts and centrocytes; two types of CD4$^+$ T cells, antigen specific variable α and β TCR repertoire and NK-like invariable α limited β TCR repertoire; as well as two types of dendritic cells, one the follicular dendritic cell and the other resembling interdigitating dendritic cells. The high rate of apoptosis that occurs in GCs is thought to be the result of processes of positive and negative selection ongoing in different compartments of GCs. The rescue of cells through high affinity interaction with antigen localized on FDCs and subsequent presentation of antigen by GC B cells to T cells may represent the positive selection, with apoptosis as the default pathway. Negative selection may occur, aimed at the prevention of autoimmunity caused by cells with mutated sIg binding to autoantigens but not receiving T help. Nonmalignant abnormalities in GC formation are seen in aging, virally infected, genetically defective and transgenic animals.

Acknowledgments

Helpful discussions with and/or personal communications from Drs. Phyllis Linton, Riccardo DallaFavera, Jorge Caamano, Thomas Feldbush, Michael Fischer, Marie Kosco-Vilbois, Ian MacLennan, Yong-Jun Liu, Moon Nahm, Garnett Kelsoe, Seth Lederman, Katherine Knight, Randy Noelle, Kiyoshi Takatsu, Takeshi Watanabe, and Jean Claude Weill are gratefully acknowledged. This work was supported by grants from the USPHS # AG-04980 and CA-14462.

References

1. De Sousa MA, Ferguson A, Parrott DM. Ecotaxis of B cells in the mouse. Adv Exp Med Biol 1973; 29:55-62.
2. Nieuwenhuis P. Follicle-center-reactions: Germinal centers or reaction centers? Acta Morphol Neerl Scand 1972; 10:397.
3. Gray D, MacLennan IC, Lane PJ. Virgin B cell recruitment and the lifespan of memory clones during antibody responses to 2,4-dinitrophenyl-hemocyanin. Eur J Immunol 1986; 16:641-648.

4. Gretz JE, Anderson AO, Shaw S. Cords, channels, corridors and conduits: Critical architectural elements facilitating cell interactions in the lymph node cortex. Immunol Rev 1997; 156:11-24.
5. Cyster JG, Goodnow CC. Antigen-induced exclusion from follicles and anergy are separate and complementary processes that influence peripheral B cell fate. Immunity 1995; 3:691-701.
6. Fulcher DA, Lyons AB, Korn SL et al. The fate of self-reactive B cells depends primarily on the degree of antigen receptor engagement and availability of T cell help. J Exp Med 1996; 183:2313-2328.
7. Thorbecke GJ, Keuning FJ. Antibody and gamma globulin formation in vitro in hemopoietic organs, antibody and gamma globulin formation. J Infect Dis 1956; 98:157-171.
8. Thorbecke GJ, Asofsky RM, Hochwald GM et al. Gamma globulin and antibody formation in vitro. III. Induction of secondary response at different intervals after the primary: The role of secondary nodules in the preparation for the secondary response. J Exp Med 1962; 116:295-309.
9. Jacob J, Kassir R, Kelsoe G. In situ studies of the primary immune response to (4-hydroxy-3-nitrophenyl)acetyl. I. The architecture and dynamics of responding cell populations. J Exp Med 1991; 173: 1165-1175.
10. Langevoort HL, Asofsky RM, Jacobson EB et al. Gamma globulin and antibody formation in vitro II. Parallel observations on histologic changes and antibody formation in the white and red pulp of the rabbit spleen during the primary response with special reference to the effect of endotoxin. J Immunol 1963; 90:60.
11. Leduc EH, Coons AH, Connolly JM. Studies on antibody production. II. The primary and secondary responses in the popliteal lymph node of the rabbit. J Exp Med 1955; 169:61-71.
12. Toellner KM, Gulbranson-Judge A, Taylor DR et al. Immunoglobulin switch transcript production in vivo related to the site and time of antigen-specific B cell activation. J Exp Med 1996; 183:2303-2312.
13. Buerki H, Kraft R, Hess MW et al. Germinal center kinetics in lymph nodes of primed mice stimulated with complexed as opposed to free antigen. Immunol Lett 1989; 23:87-94.
14. Liu YJ, Zhang J, Lane PJ et al. Sites of specific B cell activation in primary and secondary responses to T cell-dependent and T cell-independent antigens. Eur J Immunol 1991; 21:2951-2962.
15. Kroese FG, Wubbena AS, Seijen HG et al. Germinal centers develop oligoclonally. Eur J Immunol 1987; 17:1069-1072.
16. Jacob J, Kelsoe G, Rajewsky K et al. Intraclonal generation of antibody mutants in germinal centres. Nature 1991; 354:389-392.
17. MacLennan IC, Liu YJ, Johnson GD. Maturation and dispersal of B-cell clones during T cell-dependent antibody responses. Immunol Rev 1992; 126:143-161.
18. Hardie DL, Johnson GD, Khan M et al. Quantitative analysis of molecules which distinguish functional compartments within germinal centers. Eur J Immunol 1993; 23:997-1004.

19. MacLennan IC, Johnson GD, Liu YJ et al. The heterogeneity of follicular reactions. Res Immunol 1991; 142:253-257.
20. Brachtel EF, Washiyama M, Johnson GD et al. Differences in the germinal centres of palatine tonsils and lymph nodes. Scand J Immunol 1996; 43:239-247.
21. Kosco-Vilbois MH, Zentgraf H, Gerdes J et al. To 'B' or not to 'B' a germinal center? Immunol Today 1997; 18:225-230.
22. Fliedner TM, Kesse M, Gonkite EP et al. Cell proliferation in germinal centers of the rat spleen. NY Acad Sci 1964; 113:578-595.
23. Rose ML, Birbeck MS, Wallis VJ et al. Peanut lectin binding properties of germinal centres of mouse lymphoid tissue. Nature 1980; 284:364-366.
24. Weinstein PD, Anderson AO, Mage RG. Rabbit IgH sequences in appendix germinal centers: VH diversification by gene conversion-like and hypermutation mechanisms. Immunity 1994; 1:647-659.
25. Coico RF, Bhogal BS, Thorbecke GJ. Relationship of germinal centers in lymphoid tissue to immunologic memory. VI. Transfer of B cell memory with lymph node cells fractionated according to their receptors for peanut agglutinin. J Immunol 1983; 131:2254-2257.
26. Berman MA, Rafiei S, Gutman GA. Association of T cells with proliferating cells in lymphoid follicles. Transplantation 1981; 32:426-430.
27. Swerdlow SH, Murray LJ. Natural killer (Leu 7+) cells in reactive lymphoid tissues and malignant lymphomas. Am J Clin Pathol 1984; 81:459-463.
28. Liu YJ, Johnson GD, Gordon J et al. Germinal centres in T-cell-dependent antibody responses. Immunol Today 1992; 13:17-21.
29. Linton PJ, Lo D, Lai L et al. Among naive precursor cell subpopulations only progenitors of memory B cells originate germinal centers. Eur J Immunol 1992; 22:1293-1297.
30. Odartchenko N, Lewerenz M, Sordat B et al. Kinetics of cellular death in germinal centers in mouse spleen. In: Cottier H, Odartchenko N, Schindler R, Congdon CC, Eds. Germinal Centers in Immune Response. Vol. 71. Springer-Verlag, Berlin, 1967.
31. Butcher EC, Rouse RV, Coffman RL et al. Surface phenotype of Peyer's patch germinal center cells: Implications for the role of germinal centers in B cell differentiation. J Immunol 1982; 129: 2698-2707.
32. Kraal G, Weissman IL, Butcher EC. Germinal centre B cells: Antigen specificity and changes in heavy chain class expression. Nature 1982; 298:377-379.
33. Lebman DA, Griffin PM, Cebra JJ. Relationship between expression of IgA by Peyer's patch cells and functional IgA memory cells. J Exp Med 1987; 166:1405-1418.
34. Kroese FG, Wubbena AS, Opstelten D et al. B lymphocyte differentiation in the rat: Production and characterization of monoclonal antibodies to B lineage-associated antigens. Eur J Immunol 1987; 17:921-928.

35. Poppema S, Hollema H, Visser L et al. Monoclonal antibodies (MT1, MT2, MB1, MB2, MB3) reactive with leukocyte subsets in paraffin-embedded tissue sections. Am J Pathol 1987; 127:418-429.

36. Weinberg DS, Ault KA, Gurley M et al. The human lymph node germinal center cell: Characterization and isolation by using two-color flow cytometry. J Immunol 1986; 137:1486-1494.

37. Gregory CD, Tursz T, Edwards CF et al. Identification of a subset of normal B cells with a Burkitt's lymphoma (BL)-like phenotype. J Immunol 1987; 139:313-318.

38. Mangeney M, Richard Y, Coulaud D et al. CD77: An antigen of germinal center B cells entering apoptosis. Eur J Immunol 1991; 21:1131-1140.

39. Reichert RA, Gallatin WM, Weissman IL et al. Germinal center B cells lack homing receptors necessary for normal lymphocyte recirculation. J Exp Med 1983; 157:813-827.

40. Kansas GS, Wood GS, Engleman EG. Maturational and functional diversity of human B lymphocytes delineated with anti-Leu-8. J Immunol 1985; 134:3003-3006.

41. Kraal G, Hardy RR, Gallatin WM et al. Antigen-induced changes in B cell subsets in lymph nodes: Analysis by dual fluorescence flow cytofluorometry. Eur J Immunol 1986; 16:829-834.

42. Kraal G, Weissman IL, Butcher EC. Memory B cells express a phenotype consistent with migratory competence after secondary but not short-term primary immunization. Cell Immunol 1988; 115:78-87.

43. De Luaces CP, Peral JI, Tejeiro MG et al. Immunophenotypic characterization of primary and secondary lymphoid follicles. Histol Histopath 1988; 3:69-80.

44. Suzuki T, Sanders SK, Butler JL et al. Identification of an early activation antigen (Bac-1) on human B cells. J Immunol 1986; 137:1208-1213.

45. De Saint-Vis B, Cupillard L, Pandrau-Garcia D et al. Distribution of carboxypeptidase-M on lymphoid and myeloid cells parallels the other zinc-dependent proteases CD10 and CD13. Blood 1995; 86:1098-1105.

46. Prieto J, Takei F, Gendelman R et al. MALA-2, mouse homologue of human adhesion molecule ICAM-1 (CD54). Eur J Immunol 1989; 19:1551-1557.

47. Liu YJ, Malisan F, de Bouteiller O et al. Within germinal centers, isotype switching of immunoglobulin genes occurs after the onset of somatic mutation. Immunity 1996; 4:241-250.

48. Pascual V, Liu YJ, Magalski A et al. Analysis of somatic mutation in five B cell subsets of human tonsil. J Exp Med 1994; 180:329-339.

49. Dono M, Zupo S, Masante R et al. Identification of two distinct CD5⁻ B cell subsets from human tonsils with different responses to CD40 monoclonal antibody. Eur J Immunol 1993; 23:873-881.

50. Oliver AM, Martin F, Kearney JF. Mouse CD38 is down-regulated on germinal center B cells and mature plasma cells. J Immunol 1997; 158:1108-1115.

51. Kalled SL, Siva N, Stein H et al. The distribution of CD10 (NEP 24.11, CALLA) in humans and mice is similar in non-lymphoid organs but differs within the hematopoietic system: Absence on murine T and B lymphoid progenitors. Eur J Immunol 1995; 25:677-687.

52. Nahm MH, Takes PA, Bowen MB et al. Subpopulations of B lymphocytes in germinal centers, II. A germinal center B cell subpopulation expresses sIgD and CD23. Immunol Lett 1989; 21:201-208.

53. Camp RL, Kraus TA, Birkeland ML et al. High levels of CD44 expression distinguish virgin from antigen-primed B cells. J Exp Med 1991; 173:763-766.

54. Feuillard J, Taylor D, Casamayor-Palleja M et al. Isolation and characteristics of tonsil centroblasts with reference to Ig class switching. Int Immunol 1995; 7:121-130.

55. Vandenberghe P, Delabie J, de Boer M et al. In situ expression of B7/BB1 on antigen-presenting cells and activated B cells: An immunohistochemical study. Int Immunol 1993; 5:317-321.

56. Vyth-Dreese FA, Dellemijn TA, Majoor D et al. Localization in situ of the co-stimulatory molecules B7.1, B7.2, CD40 and their ligands in normal human lymphoid tissue. Eur J Immunol 1995; 25:3023-3029.

57. Sakthivel R, Christensson B, Ehlin-Henriksson B et al. Immunophenotypic characterization of follicle-center-cell-derived non-Hodgkin's lymphomas. Int J Cancer 1989; 43:624-630.

58. Fyfe G, Cebra-Thomas JA, Mustain E et al. Subpopulations of B lymphocytes in germinal centers. J Immunol 1987; 139:2187-2194.

59. Kalisiak A, Minniti JG, Oosterwijk E et al. Neutral glycosphingolipid expression in B-cell neoplasms. Int J Cancer 1991; 49:837-845.

60. Madassery JV, Gillard B, Marcus DM et al. Subpopulations of B cells in germinal centers. III. HJ6, a monoclonal antibody, binds globoside and a subpopulation of germinal center B cells. J Immunol 1991; 147:823-829.

61. To SS, Magoulas T, Nicholson E et al. Identification of a human endothelial cell activation antigen that is co-expressed by germinal follicle centre B lymphocytes. Immunology 1992; 76:616-624.

62. Epstein AL, Marder RJ, Winter JN et al. Two new monoclonal antibodies (LN-1, LN-2) reactive in B5 formalin-fixed, paraffin-embedded tissues with follicular center and mantle zone human B lymphocytes and derived tumors. J Immunol 1984; 133:1028-1036.

63. Pulford K, Micklem KJ, Jones M et al. A novel internal antigen which distinguishes germinal centre cells from other B-cell types. Immunology 1994; 82:154-163.

64. Hockenbery DM, Zutter M, Hickey W et al. BCL2 protein is topographically restricted in tissues characterized by apoptotic cell death. Proc Natl Acad Sci USA 1991; 88:6961-6965.

65. Yoshino T, Kondo E, Cao L et al. Inverse expression of bcl-2 protein and Fas antigen in lymphoblasts in peripheral lymph nodes and activated peripheral blood T and B lymphocytes. Blood 1994; 83:1856-1861.

66. Mandik L, Nguyen KA, Erikson J. Fas receptor expression on B-lineage cells. Eur J Immunol 1995; 25:3148-3154.
67. Smith KG, Nossal GJ, Tarlinton DM. Fas is highly expressed in the germinal center but is not required for regulation of the B-cell response to antigen. Proc Natl Acad Sci USA 1995; 92:11628-11632.
68. Krajewski S, Bodrug S, Gascoyne R et al. Immunohistochemical analysis of Mcl-1 and Bcl-2 proteins in normal and neoplastic lymph nodes. Am J Pathol 1994; 145:515-525.
69. Martinez-Valdez H, Guret C, de Bouteiller O et al. Human germinal center B cells express the apoptosis-inducing genes Fas, c-myc, P53, and Bax but not the survival gene bcl-2. J Exp Med 1996; 183:971-977.
70. Krajewski S, Krajewska M, Shabaik A et al. Immunohistochemical determination of in vivo distribution of Bax, a dominant inhibitor of Bcl-2. Am J Pathol 1994; 145:1323-1336.
71. Han S, Zheng B, Schatz DG et al. Neoteny in lymphocytes: Rag1 and Rag2 expression in germinal center B cells. Science 1996; 274:2094-2097.
72. Hikida M, Mori M, Kawabata T et al. Characterization of B cells expressing recombination activating genes in germinal centers of immunized mouse lymph nodes. J Immunol 1997; 158:2509-2512.
73. Hikida M, Mori M, Takai T et al. Reexpression of RAG-1 and RAG-2 genes in activated mature mouse B cells. Science 1996; 274:2092-2094.
74. Hu BT, Lee SC, Marin E et al. Telomerase is up-regulated in human germinal center B cells in vivo and can be re-expressed in memory B cells activated in vitro. J Immunol 1997; 159:1068-1071.
75. Weng N, Granger L, Hodes RJ. Telomere lengthening and telomerase activation during human B cell differentiation. Proc Natl Acad Sci USA 1997; 94:10827-10832.
75a. Kuo FC, Sklar J. Augmented expression of a human gene for 8-oxoguanine DNA glycosylase (MutM) in B lymphocytes of the dark zone in lymph node germinal centers. J Exp Med 1997; 186:1547-1556.
76. Cattoretti G, Chang CC, Cechova K et al. Bcl-6 protein is expressed in germinal-center B cells. Blood 1995; 86:45-53.
77. Onizuka T, Moriyama M, Yamochi T et al. Bcl-6 gene product, a 92- to 98-kD nuclear phosphoprotein, is highly expressed in germinal center B cells and their neoplastic counterparts. Blood 1995; 86:28-37.
78. Flenghi L, Ye BH, Fizzotti M et al. A specific monoclonal antibody (PG-B6) detects expression of the Bcl-6 protein in germinal center B cells. Am J Pathol 1995; 147:405-411.
79. Migliazza A, Martinotti S, Chen W et al. Frequent somatic hypermutation of the 5' noncoding region of the BCL6 gene in B-cell lymphoma. Proc Natl Acad Sci USA 1995; 92:12520-12524.
80. Allman D, Jain A, Dent A et al. Bcl-6 expression during B-cell activation. Blood 1996; 87:5257-5268.
81. Falini B, Fizzotti M, Pileri S et al. Bcl-6 protein expression in normal and neoplastic lymphoid tissues. Ann Oncol 1997; 8 Suppl 2:101-104.

82. Su GH, Ip HS, Cobb BS et al. The Ets protein Spi-B is expressed exclusively in B cells and T cells during development. J Exp Med 1996; 184:203-214.

83. Golay J, Erba E, Bernasconi S et al. The A-myb gene is preferentially expressed in tonsillar CD38$^+$, CD39$^-$, and sIgM$^-$ B lymphocytes and in Burkitt's lymphoma cell lines. J Immunol 1994; 153:543-553.

84. Christoph T, Rickert R, Rajewsky K. M17: A novel gene expressed in germinal centers. Int Immunol 1994; 6:1203-1211.

85. Katz P, Whalen G, Kehrl JH. Differential expression of a novel protein kinase in human B lymphocytes. Preferential localization in the germinal center. J Biol Chem 1994; 269:16802-16809.

86. Pombo CM, Kehrl JH, Sanchez I et al. Activation of the SAPK pathway by the human STE20 homologue germinal centre kinase. Nature 1995; 377:750-754.

87. Nawa MK, Fujii H, Fukumoto T et al. Involvement of thymus cells in the formation of germinal centers. Acta Anat (Basel) 1974; 90:585-590.

88. Rozing J, Brons NH, van Ewijk W et al. B lymphocyte differentiation in lethally irradiated and reconstituted mice. A histological study using immunofluorescent detection of B lymphocytes. Cell Tissue Res 1978; 189:19-30.

89. Gray D. Recruitment of virgin B cells into an immune response is restricted to activation outside lymphoid follicles. Immunology 1988; 65:73-79.

90. Durkin HG, Theis GA, Thorbecke GJ. Bursa of fabricius as site of origin of germinal centre cells. Nat New Biol 1972; 235:118-119.

91. deKruyff RH, Onikul SR, Thorbecke GJ. Migratory patterns of B lymphocytes. V. Surface Ig and migration properties of density gradient-separated bursa cells. Eur J Immunol 1976; 6:462-467.

92. Nieuwenhuis P, Keuning FJ. Germinal centres and the origin of the B-cell system. II. Germinal centres in the rabbit spleen and popliteal lymph nodes. Immunology 1974; 26:509-519.

93. Opstelten D, Deenen GJ, Stikker R et al. Germinal centers and the B cell system. VIII. Functional characteristics and cell surface markers of germinal center cell subsets differing in density and in sedimentation velocity. Immunobiology 1983; 165:1-14.

94. Vonderheide RH, Hunt SV. Surface IgD phenotype of rat germinal centre precursor cells. Adv Exp Med Biol 1988; 237:239-243.

95. Vonderheide RH, Hunt SV. Comparison of IgD$^+$ and IgD$^-$ thoracic duct B lymphocytes as germinal center precursor cells in the rat. Int Immunol 1991; 3:1273-1281.

96. Seijen HG, Bun JC, Wubbena AS et al. The germinal center precursor cell is surface mu and delta positive. Adv Exp Med Biol 1988; 237:233-237.

97. Swenson CD, Van Vollenhoven RF, Xue B et al. Physiology of IgD. IX. Effect of IgD on immunoglobulin production in young and old mice. Eur J Immunol 1988; 18:13-20.

98. Coico RF, Siskind GW, Thorbecke GJ. Role of IgD and T delta cells in the regulation of the humoral immune response. Immunol Rev 1988; 105:45-67.

99. Swenson CD, Rizinashvili E, Amin AR et al. Oligomeric IgD augments and monomeric IgD inhibits the generation of IgG memory antibody responses in normal, but not in IgD-deficient, mice. J Immunol 1995; 154:653-663.

100. Jacobson EB, Baine Y, Chen YW et al. Physiology of IgD. I. Compensatory phenomena in B lymphocyte activation in mice treated with anti-IgD antibodies. J Exp Med 1981; 154:318-332.

101. Thorbecke GJ, Flotte TJ, Baine Y. Maturity of precursor cells for germinal centers. Adv Exp Med Biol 1982; 149:845-847.

102. Nitschke L, Kosco MH, Kohler G et al. Immunoglobulin D-deficient mice can mount normal immune responses to thymus-independent and -dependent antigens. Proc Natl Acad Sci USA 1993; 90:1887-1891.

103. Roes J, Rajewsky K. Immunoglobulin D (IgD)-deficient mice reveal an auxiliary receptor function for IgD in antigen-mediated recruitment of B cells. J Exp Med 1993; 177:45-55.

104. Kroese FG, Seijen HG, Nieuwenhuis P. The initiation of germinal centre reactivity. Res Immunol 1991; 142:249-252.

105. Linton PL, Decker DJ, Klinman NR. Primary antibody-forming cells and secondary B cells are generated from separate precursor cell subpopulations. Cell 1989; 59:1049-1059.

106. Hahne M, Wenger RH, Vestweber D et al. The heat-stable antigen can alter very late antigen 4-mediated adhesion. J Exp Med 1994; 179:1391-1395.

107. Fulcher DA, Basten A. Influences on the lifespan of B cell subpopulations defined by different phenotypes. Eur J Immunol 1997; 27:1188-1199.

108. Jacobson EB, Caporale LH, Thorbecke GJ. Effect of thymus cell injections on germinal center formation in lymphoid tissues of nude (thymusless) mice. Cell Immunol 1974; 13:416-430.

109. Stedra J, Cerny J. Distinct pathways of B cell differentiation. I. Residual T cells in athymic mice support the development of splenic germinal centers and B cell memory without an induction of antibody. J Immunol 1994; 152:1718-1726.

110. De Sousa M, Pritchard H. The cellular basis of immunological recovery in nude mice after thymus grafting. Immunology 1974; 26:769-776.

111. Vonderheide RH, Hunt SV. Immigration of thoracic duct B lymphocytes into established germinal centers in the rat. Eur J Immunol 1990; 20:79-86.

112. Fuller KA, Kanagawa O, Nahm MH. T cells within germinal centers are specific for the immunizing antigen. J Immunol 1993; 151: 4505-4512.

113. Zheng B, Xue W, Kelsoe G. Locus-specific somatic hypermutation in germinal centre T cells. Nature 1994; 372:556-559.

114. Gulbranson-Judge A, MacLennan I. Sequential antigen-specific growth of T cells in the T zones and follicles in response to pigeon cytochrome C. Eur J Immunol 1996; 26:1830-1837.

115. Poppema S, Bhan AK, Reinherz EL et al. Distribution of T cell subsets in human lymph nodes. J Exp Med 1981; 153:30-41.

116. Porwit-Ksiazek A, Ksiazek T, Biberfeld P. Leu 7+ (HNK-1+) cells. I. Selective compartmentalization of Leu 7+ cells with different immunophenotypes in lymphatic tissues and blood. Scand J Immunol 1983; 18:485-493.

117. Damle NK, Doyle LV, Bradley EC. Interleukin 2-activated human killer cells are derived from phenotypically heterogeneous precursors. J Immunol 1986; 137:2814-2822.

118. Janossy G, Bofill M, Rowe D et al. The tissue distribution of T lymphocytes expressing different CD45 polypeptides. Immunology 1989; 66:517-525.

119. Pape KA, Khoruts A, Mondino A et al. Inflammatory cytokines enhance the in vivo clonal expansion and differentiation of antigen-activated CD4$^+$ T cells. J Immunol 1997; 159:591-598.

120. Pape KA, Kearney ER, Khoruts A et al. Use of adoptive transfer of T-cell-antigen-receptor-transgenic T cell for the study of T-cell activation in vivo. Immunol Rev 1997; 156:67-78.

121. Bowen MB, Butch AW, Parvin CA et al. Germinal center T cells are distinct helper-inducer T cells. Hum Immunol 1991; 31:67-75.

122. Agematsu K, Kobata T, Sugita K et al. Direct cellular communications between CD45Ro and CD45RA T cell subsets via CD27/CD70. J Immunol 1995; 154:3627-3635.

123. Agematsu K, Nagumo H, Yang FC et al. B cell subpopulations separated by CD27 and crucial collaboration of CD27$^+$ B cells and helper T cells in immunoglobulin production. Eur J Immunol 1997; 27:2073-2079.

124. Rouse RV, Ledbetter JA, Weissman IL. Mouse lymph node germinal centers contain a selected subset of T cells—the helper phenotype. J Immunol 1982; 128:2243-2246.

125. Vonderheide RH, Hunt SV. Does the availability of either B cells or CD4$^+$ cells limit germinal centre formation? Immunology 1990; 69:487-489.

126. Lederman S, Yellin MJ, Inghirami G et al. Molecular interactions mediating T-B lymphocyte collaboration in human lymphoid follicles. Roles of T cell-B-cell-activating molecule (5c8 antigen) and CD40 in contact-dependent help. J Immunol 1992; 149:3817-3826.

127. Castan J, Tenner-Racz K, Racz P et al. Accumulation of CTLA-4 expressing T lymphocytes in the germinal centres of human lymphoid tissues. Immunology 1997; 90:265-271.

128. Facchetti F, Appiani C, Salvi L et al. Immunohistologic analysis of ineffective CD40-CD40 ligand interaction in lymphoid tissues from patients with X-linked immunodeficiency with hyper-IgM. Abortive germinal center cell reaction and severe depletion of follicular dendritic cells. J Immunol 1995; 1545:6624-6633.

129. Andersson E, Ohlin M, Borrebaeck CA et al. CD4$^+$CD57$^+$ T cells derived from peripheral blood do not support immunoglobulin production by B cells. Cell Immunol 1995; 163:245-253.

130. Cardell S, Tangri S, Chan S et al. CD1-restricted CD4$^+$ T cells in major histocompatibility complex class II-deficient mice. J Exp Med 1995; 182:993-1004.

131. Cosgrove D, Gray D, Dierich A et al. Mice lacking MHC Class II molecules. Cell 1991; 66:1051-1066.

132. Behar SM, Porcelli SA, Beckman EM et al. A pathway of costimulation that prevents anergy in CD28$^-$ T cells: B7-independent costimulation of CD1-restricted T cells. J Exp Med 1995; 182:2007-2018.

133. Bendelac A, Lantz O, Quimby ME et al. CD1 recognition by mouse NK1+ T lymphocytes. Science 1995; 268:863-865.

134. Mendiratta SK, Martin WD, Hong S et al. CD1d1 mutant mice are deficient in natural T cells that promptly produce IL-4. Immunity 1997; 6:469-477.

135. Blumberg RS, Gerdes D, Chott A et al. Structure and function of the CD1 family of MHC-like cell surface proteins. Immunol Rev 1995; 147:5-29.

136. Exley M, Garcia J, Balk SP et al. Requirements for CD1d recognition by human invariant valpha24(+) CD4(-)CD8(-) T cells. J Exp Med 1997; 186:109-120.

137. Davodeau F, Peyrat MA, Necker A et al. Close phenotypic and functional similarities between human and murine alphabeta T cells expressing invariant TCR alpha-chains. J Immunol 1997; 158:5603-5611.

138. Brown DR, Fowell DJ, Corry DB et al. Beta 2-microglobulin-dependent NK1.1+ T cells are not essential for T helper cell 2 immune responses. J Exp Med 1996; 184:1295-1304.

139. Smiley ST, Kaplan MH, Grusby MJ. Immunoglobulin E production in the absence of interleukin-4-secreting CD1-dependent cells. Science 1997; 275:977-979.

140. Eikelenboom P, Boorsma DM, Van Rooyen N. The development of IgM- and IgG-containing plasmablasts in the white pulp of the spleen after stimulation with a thymus-independent antigen (LPS) and a thymus-dependent antigen (SRBC). An immunohistoperoxidase study. Cell Tissue Res 1982; 229:83-93.

141. Razzeca KJ, Pillemer E, Weissman IL et al. In situ identification of idiotype-positive cells participating in the immune response to phosphorylcholine. Eur J Immunol 1986; 16:393-399.

142. Wang D, Wells SM, Stall AM et al. Reaction of germinal centers in the T-cell-independent response to the bacterial polysaccharide alpha(1→6)dextran. Proc Natl Acad Sci USA 1994; 91:2502-2506.

143. Miller C, Stedra J, Kelsoe G et al. Facultative role of germinal centers and T cells in the somatic diversification of IgVH genes. J Exp Med 1995; 181:1319-1331.

144. Schrater AF, Goidl EA, Thorbecke GJ et al. Production of auto-anti-idiotypic antibody during the normal immune response to TNP-ficoll. III. Absence in nu/nu mice: Evidence for T-cell dependence of the anti-idiotypic-antibody response. J Exp Med 1979; 150:808-817.

145. Brossay L, Jullien D, Cardell S et al. Mouse CD1 is mainly expressed on hemopoietic-derived cells. J Immunol 1997; 159:1216-1224.

146. Wen L, Pao W, Wong FS et al. Germinal center formation, immunoglobulin class switching, and autoantibody production driven by "non alpha/beta" T cells. J Exp Med 1996; 183:2271-2282.

147. Tanaka Y, Morita CT, Tanaka Y et al. Natural and synthetic nonpeptide antigens recognized by human gamma delta T cells. Nature 1995; 375:155-158.

148. Tanaka Y, Brenner MB, Bloom BR et al. Recognition of nonpeptide antigens by T cells. J Mol Med 1996; 74:223-231.

149. Horner AA, Jabara H, Ramesh N et al. Gamma/delta T lymphocytes express CD40 ligand and induce isotype switching in B lymphocytes. J Exp Med 1995; 181:1239-1244.

150. Hermann P, Van-Kooten C, Gaillard C et al. CD40 ligand-positive CD8+ T cell clones allow B cell growth and differentiation. Eur J Immunol 1995; 25:2972-2977.

151. Thorbecke GJ, Gordon HA, Wostmann B et al. Lymphoid tissue and serum gamma globulin in young germfree chickens. J Infect Diseases 1957; 101:237-251.

152. Thorbecke GJ. Some histological and functional aspects of lymphoid tissues in germfree animals. I. Morphological studies. Ann NY Acad Sci 1959; 78:237-246.

153. Shroff KE, Cebra JJ. Development of mucosal humoral immune responses in germ-free (GF) mice. Adv Exp Med Biol 1995; 371A: 441-446.

154. Thorbecke GJ. Gamma globulin formation and antibody production in vitro. I. Gamma globulin formation in tissues from immature and normal adult rabbits. J Exp Med 1960; 112:279-292.

155. Kramer DR, Cebra JJ. Early appearance of "natural" mucosal IgA responses and germinal centers in suckling mice developing in the absence of maternal antibodies. J Immunol 1995; 154:2051-2062.

156. Reynaud CA, Garcia C, Hein WR et al. Hypermutation generating the sheep immunoglobulin repertoire is an antigen-independent process. Cell 1995; 80:115-125.

157. Reynolds JD, Morris B. The evolution and involution of Peyer's patches in fetal and postnatal sheep. Eur J Immunol 1983; 13:627-635.

158. Reynolds JD, Morris B. The effect of antigen on the development of Peyer's patches in sheep. Eur J Immunol 1984; 14:1-6.

159. Pernis B, Cohen MW, Thorbecke GJ. Specificity of reactions to antigenic stimulation in lymph nodes of immature rabbits. I. Morphological changes and gamma globulin production following stimulation with diphtheria toxoid and silica. J Immuol 1963; 91:541-552.

160. Cohen MW, Thorbecke GJ. Specificity of reaction to antigenic stimulation in lymph nodes of immature rabbits. II. Suppression of local morphologic reactions to alum precipitated bovine serum albumin by intraperitoneal injections of soluble bovine serum albumin in neonatal rabbits. J Immunol 1964; 93:629-636.

161. Pulendran B, Karvelas M, Nossal GJ. A form of immunologic tolerance through impairment of germinal center development. Proc Natl Acad Sci USA 1994; 91:2639-2643.

162. Cook MC, Basten A, Groth BFS. Outer periarteriolar lymphoid sheath arrest and subsequent differentiation of both naive and tolerant immunoglobulin transgenic B cells is determined by B cell receptor occupancy. J Exp Med 1997; 186:631-643.

163. Liu YJ. Sites of B lymphocyte selection, activation, and tolerance in spleen. J Exp Med 1997; 5:625-629.

164. Klaus GG, Humphrey JH, Kunkl A et al. The follicular dendritic cell: Its role in antigen presentation in the generation of immunological memory. Immunol Rev 1980; 53:3-28.

165. Tew JG, Phipps RP, Mandel TE. The maintenance and regulation of the humoral immune response: Persisting antigen and the role of follicular antigen-binding dendritic cells as accessory cells. Immunol Rev 1980; 53:175-201.

166. Nieuwenhuis P, Kroese FG, Opstelten D et al. De novo germinal center formation. Immunol Rev 1992; 126:77-98.

167. Kosco MH, Gray D. Signals involved in germinal center reactions. Immunol Rev 1992; 126:63-76.

168. Reynes M, Aubert JP, Cohen JH et al. Human follicular dendritic cells express CR1, CR2, and CR3 complement receptor antigens. J Immunol 1985; 135:2687-2694.

169. Gordon J, Flores-Romo L, Cairns JA et al. CD23: A multi-functional receptor/lymphokine? Immunol Today 1989; 10:153-157.

170. Schriever F, Freedman AS, Freeman G et al. Isolated human follicular dendritic cells display a unique antigenic phenotype. J Exp Med 1989; 169:2043-2058.

171. Liu YJ, Xu J, de Bouteiller O et al. Follicular dendritic cells specifically express the long CR2/CD21 isoform. J Exp Med 1997; 185:165-170.

172. Williams GM, Nossal GJ. Ontogeny of the immune response. I. The development of the follicular antigen-trapping mechanism. J Exp Med 1966; 124:47-56.

173. Nossal GJ, Abbot A, Mitchell J. Antigens in immunity. XIV. Electron microscopic radioautographic studies of antigen capture in the lymph node medulla. J Exp Med 1968; 127:263-276.

174. Mandel TE, Phipps RP, Abbot A et al. The follicular dendritic cell: Long term antigen retention during immunity. Immunol Rev 1980; 53:29-59.

175. Hanna MG Jr, Szakal AK, Tyndall RL. Histoproliferative effect of Rauscher leukemia virus on lymphatic tissue: Histological and ultrastructural studies of germinal centers and their relation to leukemogenesis. Cancer Res 1970; 30:1748-1763.

176. Martinez-Pomares L, Kosco-Vilbois M, Darley E et al. Fc chimeric protein containing the cysteine-rich domain of the murine mannose receptor binds to macrophages from splenic marginal zone and lymph node subcapsular sinus and to germinal centers. J Exp Med 1996; 184:1927-1937.

177. Gray D, Kumararatne DS, Lortan J et al. Relation of intra-splenic migration of marginal zone B cells to antigen localization on follicular dendritic cells. Immunology 1984; 52:659-669.
178. Heinen E, Braun M, Coulie PG et al. Transfer of immune complexes from lymphocytes to follicular dendritic cells. Eur J Immunol 1986; 16:167-172.
179. Anderson DW, DeNobile J, Zhao F et al. Sampling lymph node content of human immunodeficiency virus type 1 nucleic acids and p24 antigen by fine-needle aspiration in early-stage patients. J Acquir Immune Defic Syndr Hum Retrovirol 1995; 10 Suppl 2:S57-61.
180. Anderson GW, Rowland RR, Palmer GA et al. Lactate dehydrogenase-elevating virus replication persists in liver, spleen, lymph node, and testis tissues and results in accumulation of viral RNA in germinal centers, concomitant with polyclonal activation of B cells. J Virol 1995; 69:5177-5185.
181. Le Tourneau A, Audouin J, Diebold J et al. LAV-like viral particles in lymph node germinal centers in patients with the persistent lymphadenopathy syndrome and the acquired immunodeficiency syndrome-related complex: An ultrastructural study of 30 cases. Hum Pathol 1986; 17:1047-1053.
182. Pantaleo G, Graziosi C, Demarest JF et al. HIV infection is active and progressive in lymphoid tissue during the clinically latent stage of disease. Nature 1993; 362:355-358.
183. Schmitz J, van Lunzen J, Tenner-Racz K et al. Follicular dendritic cells retain HIV-1 particles on their plasma membrane, but are not productively infected in asymptomatic patients with follicular hyperplasia. J Immunol 1994; 153:1352-1359.
184. Imai Y, Dobashi M, Terashima K. Postnatal development of dendritic reticulum cells and their immune complex trapping ability. Histol Histopathol 1986; 1:19-26.
185. Szakal AK, Kosco MH, Tew JG. A novel in vivo follicular dendritic cell-dependent iccosome-mediated mechanism for delivery of antigen to antigen-processing cells. J Immunol 1988; 140:341-353.
186. Kosco MH, Szakal AK, Tew JG. In vivo obtained antigen presented by germinal center B cells to T cells in vitro. J Immunol 1988; 140:354-360.
187. Gray D, Kosco M, Stockinger B. Novel pathways of antigen presentation for the maintenance of memory. Int Immunol 1991; 3:141-148.
188. Monfalcone AP, Kosco MH, Szakal AK et al. Germinal center B cells and mixed leukocyte reactions. J Leukoc Biol 1989; 46:181-188.
189. Ponzio NM, Tsiagbe VK, Thorbecke GJ. Superantigens related to B cell hyperplasia. Springer Semin Immunopathol 1996; 17:285-306.
190. Luther SA, Gulbranson-Judge A, Acha-Orbea H et al. Viral superantigen drives extrafollicular and follicular B cell differentiation leading to virus-specific antibody production. J Exp Med 1997; 185: 551-562.

191. Blankson JN, Loh DY, Morse SS. Superantigens and conventional antigens induce different responses in alpha beta T-cell receptor transgenic mice. Immunology 1995; 85:57-62.

192. Baine Y, Thorbecke GJ. Induction and persistence of local B cell memory in mice. J Immunol 1982; 128:639-643.

193. Geldof AA, van der Ende MB, Langevoort HL. Lymph node involvement in a humoral immune response. Anat Rec 1984; 209:541-546.

194. Thorbecke GJ, Bell MK. The proliferative and anamnestic antibody response of rabbit lymphoid cells in vitro. II. Effect of passive antibody on immunologic memory in lymph nodes contralateral to the site of antigen injection. J Immunol 1973; 111:1043-1047.

195. Baine Y, Ponzio NM, Thorbecke GJ. Transfer of memory cells into antigen-pretreated hosts. II. Influence of localized antigen on the migration of specific memory B cells. Eur J Immunol 1981; 11: 990-996.

196. Baine Y, Ponzio NM, Thorbecke GJ. Unilateral localization of hapten-specific B memory cells in lymph node draining a footpad injection of antigen. Adv Exp Med Biol 1982; 149:167-178.

197. Ponzio NM, Chapman-Alexander JM, Thorbecke GJ. Transfer of memory cells into antigen-pretreated hosts. I. Functional detection of migration sites for antigen-specific B cells. Cell Immunol 1977; 34:79-92.

198. Schittek B, Rajewsky K. Maintenance of B-cell memory by long-lived cells generated from proliferating precursors. Nature 1990; 346: 749-751.

199. Berek C, Ziegner M. The maturation of the immune response. Immunol Today 1993; 14:400-404.

200. Knight KL, Crane MA. Generating the antibody repertoire in rabbit. Adv Immunol 1994; 56:179-218.

201. Weinstein PD, Mage RG, Anderson AO. The appendix functions as a mammalian bursal equivalent in the developing rabbit. Adv Exp Med Biol 1994; 355:249-253.

202. White RG. The relation of the cellular response in germinal or lymphocytopoietic centers of lymph nodes to the production of antibody. Vol. 25. Holub MaJ L, Ed. Prague: Czechoslovak Acad Sci, 1960.

203. Hurlimann J, Wakefield JD, Thorbecke GJ. The effects of immunosuppressant drugs administered during germinal center proliferation on preparation for a secondary antibody response in rabbits. In: Cottier H, Adartchenko N, Schindler R, Congdon CC, Eds. Germinal Centers in Immune Response. Vol. 71. Springer-Verlag, Berlin, 1967.

204. Thorbecke GJ. Germinal centers and immunological memory. In: Hanna LF-DaMG, ed. Lymphatic Tissue and Germinal Centers in Immune Response. Advances in Experimental Medicine and Biology. Vol. 5. New York: Plenum Press, 1969:83-92.

205. Durkin HG, Thorbecke GJ. Preferential destruction of germinal centers by prednisolone and x-irradiation. Lab Invest 1972; 26:53-62.

206. Grobler P, Buerki H, Cottier H et al. Cellular bases for relative radioresistance of the antibody-forming system at advanced stages of the secondary response to tetanus toxoid in mice. J Immunol 1974; 112:2154-2165.
207. Han S, Hathcock K, Zheng B et al. Cellular interaction in germinal centers. Roles of CD40 ligand and B7-2 in established germinal centers. J Immunol 1995; 155:556-567.
208. Thorbecke GJ, Asofsky R, Jacobson EB. Gamma globulin and antibody formation in vitro. IV. The effect on the secondary response of X-irradiation given at varying intervals after a primary injection of bovine gamma globulin. J Immunol 1964; 92:734-746.
209. Porter RJ. Studies on antibody formation. Effect of x-irradiation on adaptation for the secondary response of rabbits to bovine gammaglobulin. J Immunol 1960; 84:485-490.
210. Porter RJ. Prolonged suppression by X-ray of adaptation for the secondary antibody response. Proc Soc Exp Biol Med 1962; 111: 583-587.
211. Kunkl A, Klaus GG. The generation of memory cells. IV. Immunization with antigen-antibody complexes accelerates the development of B-memory cells, the formation of germinal centres and the maturation of antibody affinity in the secondary response. Immunology 1981; 43:371-378.
212. Kraft R, Buerki H, Schweizer T et al. Tetanus toxoid complexed with heterologous antibody can induce germinal centre formation and B cell memory in mice without evoking a detectable anti-toxin response. Clin Exp Immunol 1989; 76:138-143.
213. Wakefield JD, Thorbecke GJ. Relationship of germinal centers in lymphoid tissue to immunological memory. I. Evidence for the formation of small lymphocytes upon transfer of primed splenic white pulp to syngeneic mice. J Exp Med 1968; 128:153-169.
214. Wakefield JD, Thorbecke GJ. Relationship of germinal centers in lymphoid tissue to immunological memory. II. The detection of primed cells and their proliferation upon cell transfer to lethally irradiated syngeneic mice. J Exp Med 1968; 128:171-187.
215. Mitchell GF, Chan EL, Noble MS et al. Immunological memory in mice. 3. Memory to heterologous erythrocytes in both T cell and B cell populations and requirement for T cells in expression of B cell memory. Evidence using immunoglobulin allotype and mouse alloantigen theta markers with congenic mice. J Exp Med 1972; 135:165-184.
216. Mond JJ, Takahashi T, Thorbecke GJ. Surface antigens of immunocompetent cells. 3. In vitro studies of the role of B and T cells in immunological memory. J Exp Med 1972; 136:663-675.
217. Mond JJ, Caporale LH, Thorbecke GJ. Kinetics of B cell memory development during a thymus "independent" immune response. Cell Immunol 1974; 10:105-116.

218. Romano TJ, Mond JJ, Thorbecke GJ. Immunological memory function of the T and B cell types: Distribution over mouse spleen and lymph nodes. Eur J Immunol 1975; 5:211-215.
219. Koch G, Osmond DG, Julius MH et al. The mechanism of thymus-dependent antibody formation in bone marrow. J Immunol 1981; 126:1447-1451.
220. Arpin C, Bancereau J, Liu Y-J. Memory B cells are biased towards terminal differentiation: A strategy that may prevent repertoire freezing. J Exp Med 1997; 186:1-10.
221. Beagley KW, Eldridge JH, Aicher WK et al. Peyer's patch B cells with memory cell characteristics undergo terminal differentiation within 24 hours in response to interleukin-6. Cytokine 1991; 3:107-116.
222. Liu YJ, Joshua DE, Williams GT et al. Mechanism of antigen-driven selection in germinal centres. Nature 1989; 342:929-931.
223. Liu YJ, Barthelemy C, de Bouteiller O et al. The differences in survival and phenotype between centroblasts and centrocytes. Adv Exp Med Biol 1994; 355:213-218.
224. Taga S, Tetaud C, Mangeney M et al. Sequential changes in glycolipid expression during human B cell differentiation: Enzymatic bases. Biochim Biophys Acta 1995; 1254:56-65.
225. Koopman G, Reutelingsperger CP, Kuijten GA et al. Annexin V for flow cytometric detection of phosphatidylserine expression on B cells undergoing apoptosis. Blood 1994; 84:1415-1420.
226. Kim HS, Zhang X, Klyushnenkova E et al. Stimulation of germinal center B lymphocyte proliferation by an FDC-like cell line, HK. J Immunol 1995; 155:1101-1109.
227. Lindhout E, Lakeman A, de Groot C. Follicular dendritic cells inhibit apoptosis in human B lymphocytes by a rapid and irreversible blockade of preexisting endonuclease. J Exp Med 1995; 181:1985-1995.
228. Holder MJ, Wang H, Milner AE et al. Suppression of apoptosis in normal and neoplastic human B lymphocytes by CD40 ligand is independent of Bcl-2 induction. Eur J Immunol 1993; 23:2368-2371.
229. Nakayama K, Nakayama K, Dustin LB et al. T-B cell interaction inhibits spontaneous apoptosis of mature lymphocytes in Bcl-2-deficient mice. J Exp Med 1995; 182:1101-1109.
230. Swartzendruber DC. Observations on the ultrastructure of lymphatic tissue germinal centers. In: Cottier H, Odartchneko R, Schindler R, Congdon R, eds. Germinal Centers in Immune Response. Vol. 71. Berlin: Springer-Verlag, 1967.
231. Sakuma H, Kasajima T, Imai Y et al. An electron microscopic study on the reticuloendothelial cells in the lymph nodes. Acta Pathol Jpn 1981; 31:449-472.
232. Szakal AK, Gieringer RL, Kosco MH et al. Isolated follicular dendritic cells: Cytochemical antigen localization, Nomarski, SEM, and TEM morphology. J Immunol 1985; 134:1349-1359.
233. Krenacs T, van Dartel M, Lindhout E et al. Direct cell/cell communication in the lymphoid germinal center: Connexin43 gap junctions functionally couple follicular dendritic cells to each other and to B lymphocytes. Eur J Immunol 1997; 27:1489-1497.

234. Krenacs T, Rosendaal M. Immunohistological detection of gap junctions in human lymphoid tissue: Connexin43 in follicular dendritic and lymphoendothelial cells. J Histochem Cytochem 1995; 43:1125-1137.

235. Cerny A, Zinkernagel RM, Groscurth P. Development of follicular dendritic cells in lymph nodes of B-cell-depleted mice. Cell Tissue Res 1988; 254:449-454.

236. Kapasi ZF, Burton GF, Shultz LD et al. Induction of functional follicular dendritic cell development in severe combined immunodeficiency mice. Influence of B and T cells. J Immunol 1993; 150:2648-2658.

237. Kapasi ZF, Kosco-Vilbois MH, Shultz LD et al. Cellular origin of follicular dendritic cells. Adv Exp Med Biol 1994; 355:231-235.

238. Airas L, Jalkanen S. CD73 mediates adhesion of B cells to follicular dendritic cells. Blood 1996; 88:1755-1764.

239. Rice GE, Munro JM, Bevilacqua MP. Inducible cell adhesion molecule 110 (INCAM-110) is an endothelial receptor for lymphocytes. A CD11/CD18-independent adhesion mechanism. J Exp Med 1990; 171:1369-1374.

240. Freedman AS, Munro JM, Rice GE et al. Adhesion of human B cells to germinal centers in vitro involves VLA-4 and INCAM-110. Science 1990; 249:1030-1033.

241. Koopman G, Parmentier HK, Schuurman HJ et al. Adhesion of human B cells to follicular dendritic cells involves both the lymphocyte function-associated antigen 1/intercellular adhesion molecule 1 and very late antigen 4/vascular cell adhesion molecule 1 pathways. J Exp Med 1991; 173:1297-1304.

242. Butch AW, Hug BA, Nahm MH. Properties of human follicular dendritic cells purified with HJ2, a new monoclonal antibody. Cell Immunol 1994; 155:27-41.

243. Orscheschek K, Merz H, Schlegelberger B et al. An immortalized cell line with features of human follicular dendritic cells. Antigen and cytokine expression analysis. Eur J Immunol 1994; 24:2682-2690.

244. Liu YJ, Cairns JA, Holder MJ et al. Recombinant 25-kDa CD23 and interleukin 1 alpha promote the survival of germinal center B cells: Evidence for bifurcation in the development of centrocytes rescued from apoptosis. Eur J Immunol 1991; 21:1107-1114.

245. Holder MJ, Knox K, Gordon J. Factors modifying survival pathways of germinal center B cells. Glucocorticoids and transforming growth factor-beta, but not cyclosporin A or anti-CD19, block surface immunoglobulin-mediated rescue from apoptosis. Eur J Immunol 1992; 22:2725-2728.

246. Matsumoto AK, Kopicky-Burd J, Carter RH et al. Intersection of the complement and immune systems: A signal transduction complex of the B lymphocyte-containing complement receptor type 2 and CD19. J Exp Med 1991; 173:55-64.

247. Sato S, Steeber DA, Tedder TF. The CD19 signal transduction molecule is a response regulator of B-lymphocyte differentiation. Proc Natl Acad Sci USA 1995; 92:11558-11562.
248. Myers DE, Jun X, Waddick KG et al. Membrane-associated CD19-LYN complex is an endogenous p53-independent and Bcl-2-independent regulator of apoptosis in human B-lineage lymphoma cells. Proc Natl Acad Sci USA 1995; 92:9575-9579.
249. Maloney MD, Lingwood CA. CD19 has a potential CD77 (globotriaosyl ceramide)-binding site with sequence similarity to verotoxin B-subunits: Implications of molecular mimicry for B cell adhesion and enterohemorrhagic Escherichia coli pathogenesis. J Exp Med 1994; 180:191-201.
250. Gruss HJ, Dower SK. Tumor necrosis factor ligand superfamily: Involvement in the pathology of malignant lymphomas. Blood 1995; 85:3378-3404.
251. Worm M, Geha RS. Activation of tumor necrosis factor-alpha and lymphotoxin-alpha via anti-CD40 in human B cells. Int Arch Allergy Immunol 1995; 107:368-369.
252. Boussiotis VA, Nadler LM, Strominger JL et al. Tumor necrosis factor alpha is an autocrine growth factor for normal human B cells. Proc Natl Acad Sci USA 1994; 91:7007-7011.
253. Strobach RS, Nakamine H, Masih AS et al. Nerve growth factor receptor expression on dendritic reticulum cells in follicular lymphoid proliferations. Hum Pathol 1991; 22:481-485.
254. Pezzati P, Stanisz AM, Marshall JS et al. Expression of nerve growth factor receptor immunoreactivity on follicular dendritic cells from human mucosa associated lymphoid tissues. Immunology 1992; 76:485-490.
255. Ryffel B, Brockhaus M, Durmuller U et al. Tumor necrosis factor receptors in lymphoid tissues and lymphomas. Source and site of action of tumor necrosis factor alpha. Am J Pathol 1991; 139:7-15.
256. Torcia M, Bracci-Laudiero L, Lucibello M et al. Nerve growth factor is an autocrine survival factor for memory B lymphocytes. Cell 1996; 85:345-356.
257. Santambrogio L, Benedetti M, Chao MV et al. Nerve growth factor production by lymphocytes. J Immunol 1994; 153:4488-4495.
258. Kim HS, Zhang X, Choi YS. Activation and proliferation of follicular dendritic cell-like cells by activated T lymphocytes. J Immunol 1994; 153:2951-2961.
259. Clark EA, Grabstein KH, Shu GL. Cultured human follicular dendritic cells. Growth characteristics and interactions with B lymphocytes. J Immunol 1992; 148:3327-3335.
260. Gordon J, Katira A, Strain AJ et al. Inhibition of interleukin 4-promoted CD23 production in human B lymphocytes by transforming growth factor-β, interferons or anti-CD19 antibody is overriden on engaging CD40. Eur J Immunol 1991; 21:1917-1922.
261. Leger-Ravet MB, Peuchmaur M, Devergne O et al. Interleukin-6 gene expression in Castleman's disease. Blood 1991; 78:2923-2930.

262. Burdin N, Galibert L, Garrone P et al. Inability to produce IL-6 is a functional feature of human germinal center B lymphocytes. J Immunol 1996; 156:4107-4113.
263. Kroncke R, Loppnow H, Flad HD et al. Human follicular dendritic cells and vascular cells produce interleukin-7: A potential role for interleukin-7 in the germinal center reaction. Eur J Immunol 1996; 26:2541-2544.
264. Crompton T, Outram SV, Buckland J et al. A transgenic T cell receptor restores thymocyte differentiation in interleukin-7 receptor alpha chain-deficient mice. Eur J Immunol 1997; 27:100-104.
265. Webb LM, Foxwell BM, Feldmann M. Interleukin-7 activates human naive $CD4^+$ cells and primes for interleukin-4 production. Eur J Immunol 1997; 27:633-640.
266. Moore TA, von Freeden-Jeffry U, Murray R et al. Inhibition of gamma delta T cell development and early thymocyte maturation in IL-7-/- mice. J Immunol 1996; 157:2366-2373.
267. van der Voort R, Taher TE, Keehnen RM et al. Paracrine regulation of germinal center B cell adhesion through the c-met-hepatocyte growth factor/scatter factor pathway. J Exp Med 1997; 185:2121-2131.
268. del Pozo MA, Cabanas C, Montoya MC et al. ICAMs redistributed by chemokines to cellular uropods as a mechanism for recruitment of T lymphocytes. J Cell Biol 1997; 137:493-508.
269. Adema GJ, Hartgers F, Verstraten R et al. A dendritic-cell-derived C-C chemokine that preferentially attracts naive T cells. Nature 1997; 387:713-717.
270. Forster R, Mattis AE, Kremmer E et al. A putative chemokine receptor, BLR1, directs B cell migration to defined lymphoid organs and specific anatomic compartments of the spleen. Cell 1996; 87:1037-1047.
271. Degli-Esposti MA, Davis-Smith T, Din WS et al. Activation of the lymphotoxin beta receptor by cross-linking induces chemokine production and growth arrest in A375 melanoma cells. J Immunol 1997; 158:1756-1762.
272. Grouard G, Durand I, Filgueira L et al. Dendritic cells capable of stimulating T cells in germinal centres. Nature 1996; 384:364-367.
273. Groh V, Gadner H, Radaszkiewicz T et al. The phenotypic spectrum of histiocytosis X cells. J Invest Dermatol 1988; 90:441-447.
274. Liu YJ, Grouard G, de Bouteiller O et al. Follicular dendritic cells and germinal centers. Int Rev Cytol 1996; 166:139-179.
275. Schriever F, Freeman G, Nadler LM. Follicular dendritic cells contain a unique gene repertoire demonstrated by single-cell polymerase chain reaction. Blood 1991; 77:787-791.
276. Szakal AK, Kosco MH, Tew JG. Microanatomy of lymphoid tissue during humoral immune responses: Structure function relationships. Annu Rev Immunol 1989; 7:91-109.
277. White RG, Henderson DC, Eslami MB et al. Localization of a protein antigen in the chicken spleen. Effect of various manipulative procedures on the morphogenesis of the germinal centre. Immunology 1975; 28:1-21.

278. Caux C, Massacrier C, Vanbervliet B et al. CD34⁺ hematopoietic progenitors from human cord blood differentiate along two independent dendritic cell pathways in response to granulocyte-macrophage colony-stimulating factor plus tumor necrosis factor α: II Functional analysis. Blood 1997; 90:1458-1470.

279. Mueller CGF, Ri sson M-C, Salinas B et al. Polymerase chain reaction selects a novel disintegrin proteinase from CD40-activated germinal center dendritic cells. J Exp Med 1997; 186:655-663.

280. Moss ML, Jin SL, Milla ME et al. Cloning of a disintegrin metalloproteinase that processes precursor tumour-necrosis factor-alpha. Nature 1997; 385:733-736.

281. Black RA, Rauch CT, Kozlosky CJ et al. A metalloproteinase disintegrin that releases tumour-necrosis factor-alpha from cells. Nature 1997; 385:729-733.

282. Foy JE, Laman JD, Ledbetter JA et al. Gp39-CD40 interactions are essential for germinal center formation and the development of B cell memory. J Exp Med 1994; 1806:157-163.

283. Lederman S, Yellin MJ, Cleary AM et al. T-BAM/CD40-L on helper T lymphocytes augments lymphokine-induced B cell Ig isotype switch recombination and rescues B cells from programmed cell death. J Immunol 1994; 152:2163-2171.

284. Casamayor-Palleja M, Khan M, MacLennan IC. A subset of CD4⁺ memory T cells contains preformed CD40 ligand that is rapidly but transiently expressed on their surface after activation through the T cell receptor complex. J Exp Med 1995; 181:1293-1301.

285. Kelly KA, Bucy RP, Nahm MH. Germinal center T cells exhibit properties of memory helper T cells. Cell Immunol 1995; 163:206-214.

286. van Ewijk W, van Soest PL, van den Engh GJ. Fluorescence analysis and anatomic distribution of mouse T lymphocyte subsets defined by monoclonal antibodies to the antigens Thy-1, Lyt-1, Lyt-2, and T-200. J Immunol 1981; 127:2594-2604.

287. Zheng B, Han S, Kelsoe G. T helper cells in murine germinal centers are antigen-specific emigrants that downregulate Thy-1. J Exp Med 1996; 184:1083-1091.

288. Mosnier JF, Degott C, Marcellin P et al. The intraportal lymphoid nodule and its environment in chronic active hepatitis C: An immunohistochemical study. Hepatology 1993; 17:366-371.

289. Choe J, Kim H, Armitage R et al. The functional role of B cell antigen receptor stimulation and IL-4 in the generation of human memory B cells from germinal center B cells. J Immunol 1997; 159:3757-3766.

290. Dahlenborg K, Pound JD, Gordon J et al. Terminal differentiation of human germinal center B cells in vitro. Cell Immunol 1997; 175:141-149.

291. Xu J, Foy TM, Laman JD et al. Mice deficient for the CD40 ligand. Immunity 1994; 1:423-431.

292. Kawabe T, Naka T, Yoshida K et al. The immune responses in CD40-deficient mice: Impaired immunoglobulin class switching and germinal center formation. Immunity 1994; 1:167-178.

293. Hollenbaugh D, Ochs HD, Noelle RJ et al. The role of CD40 and its ligand in the regulation of the immune response. Immunol Rev 1994; 138:23-37.

294. Noelle RJ. The role of gp39 (CD40L) in immunity. Clin Immunol Immunopathol 1995; 76:S203-207.

295. Van den Eertwegh AJ, Noelle RJ, Roy M et al. In vivo CD40-gp39 interactions are essential for thymus-dependent humoral immunity. I. In vivo expression of CD40 ligand, cytokines, and antibody production delineates sites of cognate T-B cell interactions. J Exp Med 1993; 178:1555-1565.

296. Grewal IS, Xu J, Flavell RA. Impairment of antigen-specific T-cell priming in mice lacking CD40 ligand. Nature 1995; 378:617-620.

297. van Essen D, Kikutani H, Gray D. CD40 ligand-transduced co-stimulation of T cells in the development of helper function. Nature 1995; 378:620-623.

298. Gray D, Dullforce P, Jainandunsing S. Memory B cell development but not germinal center formation is impaired by in vivo blockade of CD40-CD40 ligand interaction. J Exp Med 1994; 180:141-155.

299. Grammer AC, Bergman MC, Miura Y et al. The CD40 ligand expressed by human B cells costimulates B cell responses. J Immunol 1995; 154:4996-5010.

300. Han S, Zheng B, Dal Porto J et al. In situ studies of the primary immune response to (4-hydroxy-3-nitrophenyl)acetyl. IV. Affinity-dependent, antigen-driven B cell apoptosis in germinal centers as a mechanism for maintaining self-tolerance. J Exp Med 1995; 182: 1635-1644.

301. Zheng B, Han S, Zhu Q et al. Alternative pathways for the selection of antigen-specific peripheral T cells. Nature 1996; 384:263-266.

302. Liu YJ, Mason DY, Johnson GD et al. Germinal center cells express bcl-2 protein after activation by signals which prevent their entry into apoptosis. Eur J Immunol 1991; 21:1905-1910.

303. Wang Z, Karras JG, Howard RG et al. Induction of bcl-x by CD40 engagement rescues sIg-induced apoptosis in murine B cells. J Immunol 1995; 155:3722-3725.

304. Zhang X, Li L, Choe J et al. Up-regulation of Bcl-xL expression protects CD40-activated human B cells from Fas-mediated apoptosis. Cell Immunol 1996; 173:149-154.

305. Choi MS, Holmann M, Atkins CJ et al. Expression of bcl-x during mouse B cell differentiation and following activation by various stimuli. Eur J Immunol 1996; 26:676-682.

306. Tuscano JM, Druey KM, Riva A et al. Bcl-x rather than Bcl-2 mediates CD40-dependent centrocyte survival in the germinal center. Blood 1996; 88:1359-1364.

307. Ning ZQ, Norton JD, Li J et al. Distinct mechanisms for rescue from apoptosis in Ramos human B cells by signaling through CD40 and interleukin-4 receptor: Role for inhibition of an early response gene, Berg36. Eur J Immunol 1996; 26:2356-2363.

308. Billian G, Mondiere P, Berard M et al. Antigen receptor-induced apoptosis of human germinal center B cells is targeted to a centrocytic subset. Eur J Immunol 1997; 27:405-414.

309. Galibert L, Burdin N, Barthelemy C et al. Negative selection of human germinal center B cells by prolonged BCR cross-linking. J Exp Med 1996; 183:2075-2085.

310. Knox KA, Gordon J. Protein tyrosine phosphorylation is mandatory for CD40-mediated rescue of germinal center B cells from apoptosis. Eur J Immunol 1993; 23:2578-2584.

311. Knox KA, Gordon J. Protein tyrosine kinases couple the surface immunoglobulin of germinal center B cells to phosphatidylinositol-dependent and -independent pathways of rescue from apoptosis. Cell Immunol 1994; 155:62-76.

312. Sarma V, Lin Z, Clark L et al. Activation of the B-cell surface receptor CD40 induces A20, a novel zinc finger protein that inhibits apoptosis. J Biol Chem 1995; 270:12343-12346.

313. Levy Y, Brouet JC. Interleukin-10 prevents spontaneous death of germinal center B cells by induction of the bcl-2 protein. J Clin Invest 1994; 93:424-428.

314. Kelly K, Knox KA. Differential regulatory effects of cAMP-elevating agents on human normal and neoplastic B cell functional response following ligation of surface immunoglobulin and CD40. Cell Immunol 1995; 166:93-102.

315. Cutrona G, Ulivi M, Fais F et al. Transfection of the c-myc oncogene into normal Epstein-Barr virus-harboring B cells results in new phenotypic and functional features resembling those of Burkitt lymphoma cells and normal centroblasts. J Exp Med 1995; 181:699-711.

316. Yoshino T, Cao L, Nishiuchi R et al. Ligation of HLA class II molecules promotes sensitivity to CD95 (Fas antigen, APO-1)-mediated apoptosis. Eur J Immunol 1995; 25:2190-2194.

317. Lagresle C, Bella C, Daniel PT et al. Regulation of germinal center B cell differentiation. Role of the human APO-1/Fas (CD95) molecule. J Immunol 1995; 154:5746-5756.

318. Cleary AM, Fortune SM, Yellin MJ et al. Opposing roles of CD95 (Fas/APO-1) and CD40 in the death and rescue of human low density tonsillar B cells. J Immunol 1995; 155:3329-3337.

319. Schattner EJ, Elkon KB, Yoo DH et al. CD40 ligation induces Apo-1/Fas expression on human B lymphocytes and facilitates apoptosis through the Apo-1/Fas pathway. J Exp Med 1995; 182:1557-1565.

320. Mangeney M, Lingwood CA, Taga S et al. Apoptosis induced in Burkitt's lymphoma cells via Gb3/CD77, a glycolipid antigen. Cancer Res 1993; 53:5314-5319.

321. Mangeney M, Rousselet G, Taga S et al. The fate of human CD77[+] germinal center B lymphocytes after rescue from apoptosis. Mol Immunol 1995; 32:333-339.

322. Lund FE, Solvason NW, Cooke MP et al. Signaling through murine CD38 is impaired in antigen receptor-unresponsive B cells. Eur J Immunol 1995; 25:1338-1345.
323. Lund FE, Yu N, Kim KM et al. Signaling through CD38 augments B cell antigen receptor (BCR) responses and is dependent on BCR expression. J Immunol 1996; 157:1455-1467.
324. Zupo S, Rugari E, Dono M et al. CD38 signaling by agonistic monoclonal antibody prevents apoptosis of human germinal center B cells. Eur J Immunol 1994; 24:1218-1222.
325. Yamashita Y, Miyake K, Kikuchi Y et al. A monoclonal antibody against a murine CD38 homologue delivers a signal to B cells for prolongation of survival and protection against apoptosis in vitro: Unresponsiveness of X-linked immunodeficient B cells. Immunology 1995; 85:248-255.
326. Silvennoinen O, Nishigaki H, Kitanaka A et al. CD38 signal transduction in human B cell precursors. Rapid induction of tyrosine phosphorylation, activation of syk tyrosine kinase, and phosphorylation of phospholipase C-gamma and phosphatidylinositol 3-kinase. J Immunol 1996; 156:100-107.
327. Deaglio S, Dianzani U, Horenstein AL et al. Human CD38 ligand. A 120-KDA protein predominantly expressed on endothelial cells. J Immunol 1996; 156:727-734.
328. Butch AW, Chung GH, Hoffmann JW et al. Cytokine expression by germinal center cells. J Immunol 1993; 150:39-47.
329. Corcione A, Baldi L, Zupo S et al. Spontaneous production of granulocyte colony-stimulating factor in vitro by human B-lineage lymphocytes is a distinctive marker of germinal center cells. J Immunol 1994; 153:2868-2877.
330. Finke J, Ternes P, Lange W et al. Expression of interleukin 10 in B lymphocytes of different origin. Leukemia 1993; 7:1852-1857.
331. Bogen SA, Fogelman I, Abbas AK. Analysis of IL-2, IL-4, and IFN-gamma producing cells in situ during immune responses to protein antigen. J Immunol 1993;4197-4205.
332. Butch AW, Nahm MH. Functional properties of human germinal center B cells. Cell Immunol 1992; 140:331-344.
333. Lagresle C, Bella C, Defrance T. Phenotypic and functional heterogeneity of the IgD⁻ B cell compartment: Identification of two major tonsillar B cell subsets. Int Immunol 1993; 5:1259-1268.
334. Pound JD, Gordon J. Maintenance of human germinal center B cells in vitro. Blood 1997; 89:919-928.
335. Cocks G, de Waal Malefyt R, Galizzi JP et al. IL-13 induces proliferation and differentiation of human B cells activated by the CD40 ligand. Int Immunol 1993; 5:657-663.
336. Defrance T, Carayon P, Billian G et al. Interleukin 13 is a B cell stimulating factor. J Exp Med 1994; 179:135-143.
337. Fior R, Vita N, Raphael M et al. Interleukin-13 gene expression by malignant and EBV-transformed human B lymphocytes. Eur Cytokine Netw 1994; 5:593-600.

338. Billard C, Caput D, Vita N et al. Interleukin-13 responsiveness and interleukin-13 receptor expression in non-Hodgkin's lymphoma and reactive lymph node B cells. Modulation by CD40 activation. Eur Cytokine Netw 1997; 8:19-27.

339. Rabinowitz JL, Tsiagbe VK, Nicknam MH et al. Germinal center cells are a major IL-5-responsive B cell population in peripheral lymph nodes engaged in the immune response. J Immunol 1990; 145: 2440-2447.

340. Tsiagbe VK, Linton PJ, Thorbecke GJ. The path of memory B-cell development. Immunol Rev 1992; 126:113-141.

341. Tsiagbe VK, Chickramane S, Amin AR et al. Cytokine responsiveness of germinal center B cells. Adv Exp Med Biol 1994; 355:207-212.

342. Rosen J, Sherman WT, Farrington GL et al. Chondroitin sulfates used as adjuvants for antibody formation and as mitogens for B cells. J Biol Response Mod 1987; 6:355-366.

343. Clutterbuck E, Shields JG, Gordon J et al. Recombinant human interleukin 5 is an eosinophil differentiation factor but has no activity in standard human B cell growth factor assays. Eur J Immunol 1987; 17:1743-1750.

344. Secord EA, Rizzo LV, Barroso EW et al. Reconstitution of germinal center formation in nude mice with Th1 and Th2 clones. Cell Immunol 1996; 174:173-179.

345. Rizzo LV, DeKruyff RH, Umetsu DT. Generation of B cell memory and affinity maturation. Induction with Th1 and Th2 T cell clones. J Immunol 1992; 148:3733-3739.

346. Rizzo LV, DeKruyff RH, Umetsu DT et al. Regulation of the interaction between Th1 and Th2 T cell clones to provide help for antibody production in vivo. Eur J Immunol 1995; 25:708-716.

347. Kehrl JH, Alvarez-Mon M, Delsing GA et al. Lymphotoxin is an important T cell-derived growth factor for human B cells. Science 1987; 238:1144-1146.

348. Kehrl JH, Miller A, Fauci AS. Effect of tumor necrosis factor alpha on mitogen-activated human B cells. J Exp Med 1987; 166:786-791.

349. Zola H, Nikoloutsopoulos A. Effect of recombinant human tumour necrosis factor beta (TNF beta) on activation, proliferation and differentiation of human B lymphocytes. Immunology 1989; 67:231-236.

350. Rodriguez C, Roldan E, Navas G et al. Essential role of tumor necrosis factor-alpha in the differentiation of human tonsil in vivo induced B cells capable of spontaneous and high-rate immunoglobulin secretion. Eur J Immunol 1993; 23:1160-1164.

351. Del Prete G, De Carli M, MM DE et al. Polyclonal B cell activation induced by herpesvirus saimiri-transformed human CD4+ T cell clones. Role for membrane TNF-alpha/TNF-alpha receptors and CD2/CD58 interactions. J Immunol 1994; 153:4872-4879.

352. Vajdy M, Kosco-Vilbois MH, Kopf M et al. Impaired mucosal immune responses in interleukin 4-targeted mice. J Exp Med 1995; 181:41-53.

353. O'Shea J. Jaks, STATs, cytokine signal transduction, and immuno-regulation: Are we there yet? Immunity 1997; 7:1-11.
354. Kopf M, Le Gros G, Coyle AJ et al. Immune responses of IL-4, IL-5, IL-6 deficient mice. Immunol Rev 1995; 148:45-69.
355. Suematsu S, Matsuda T, Aozasa K et al. IgG1 plasmacytosis in interleukin 6 transgenic mice. Proc Natl Acad Sci USA 1989; 86:7547-7551.
356. Velardi A, Tilden AB, Millo R et al. Isolation and characterization of Leu 7+ germinal-center cells with the T helper-cell phenotype and granular lymphocyte morphology. J Clin Immunol 1986; 6:205-215.
357. Velardi A, Mingari MC, Moretta L et al. Functional analysis of cloned germinal center CD4$^+$ cells with natural killer cell-related features. Divergence from typical T helper cells. J Immunol 1986; 137: 2808-2813.
358. Andersson E, Dahlenborg K, Ohlin M et al. Immunoglobulin production induced by CD57$^+$ GC-derived helper T cells in vitro requires addition of exogenous IL-2. Cell Immunol 1996; 169:166-173.
359. Arase H, Arase N, Nakagawa K et al. NK1.1+ CD4$^+$ CD8- thymocytes with specific lymphokine secretion. Eur J Immunol 1993; 23:307-310.
360. Emoto M, Emoto Y, Kaufmann SHE. IL-4 producing CD4$^+$ TCR alpha-β^{int} liver lymphocytes: Influence of thymus, B$_2$-microglobulin and NK1.1 expression. Intern Immunol 1995; 7:1729-1739.
361. Yoshimoto T, Bendelac A, Watson C et al. Role of NK1.1+ T cells in a TH2 response and in immunoglobulin E production. Science 1995; 270:1845-1847.
362. Yoshimoto T, Paul WE. CD4$^+$, NK1.1$^+$ T cells promptly produce interleukin 4 in response to in vivo challenge with anti-CD3. J Exp Med 1994; 179:1285-1295.
363. Barrett TB, Shu G, Clark EA. CD40 signaling activates CD11a/CD18 (LFA-1)-mediated adhesion in B cells. J Immunol 1991; 146:1722-1729.
364. Klaus GG, Holman M, Hasbold J. Properties of mouse CD40: The role of homotypic adhesion in the activation of B cells via CD40. Eur J Immunol 1994; 24:2714-2719.
365. Roy M, Aruffo A, Ledbetter J et al. Studies on the interdependence of gp39 and B7 expression and function during antigen-specific immune responses. Eur J Immunol 1995; 25:596-603.
366. Hasbold J, Johnson-Leger C, Atkins CJ et al. Properties of mouse CD40: Cellular distribution of CD40 and B cell activation by monoclonal anti-mouse CD40 antibodies. Eur J Immunol 1994; 24: 1835-1842.
367. Lahvis GP, Cerny J. Induction of germinal center B cell markers in vitro by activated CD4$^+$ T lymphocytes: The role of CD40 ligand, soluble factors, and B cell antigen receptor cross-linking. J Immunol 1997; 159:1783-1793.
368. Wortis HH, Teutsch M, Higer M et al. B-cell activation by crosslinking of surface IgM or ligation of CD40 involves alternative signal pathways and results in different B-cell phenotypes. Proc Natl Acad Sci USA 1995; 92:3348-3352.

369. Arpin C, Dechanet J, Van Kooten C et al. Generation of memory B cells and plasma cells in vitro. Science 1995; 268:720-722.

370. Callard RE, Herbert J, Smith SH et al. CD40 cross-linking inhibits specific antibody production by human B cells. Intern Immunol 1995; 7:1809-1815.

371. Grouard G, de Bouteiller O, Barthelemy C et al. Regulation of human B cell activation by follicular dendritic cell and T cell signals. Curr Top Microbiol Immunol 1995; 201:105-117.

372. Grouard G, de Bouteiller O, Banchereau J et al. Human follicular dendritic cells enhance cytokine-dependent growth and differentiation of CD40-activated B cells. J Immunol 1995; 155:3345-3352.

373. Galibert L, Burdin N, de Saint-Vis B et al. CD40 and B cell antigen receptor dual triggering of resting B lymphocytes turns on a partial germinal center phenotype. J Exp Med 1996; 183:77-85.

374. Wheeler K, Gordon J. Co-ligation of surface IgM and CD40 on naive B lymphocytes generates a blast population with an ambiguous extrafollicular/germinal centre cell phenotype. Int Immunol 1996; 8:815-828.

375. Grimaitre M, Werner-Favre C, Kindler V et al. Human naive B cells cultured with EL-4 T cells mimic a germinal center-related B cell stage before generating plasma cells. Concordant changes in Bcl-2 protein and messenger RNA levels. Eur J Immunol 1997; 27:199-205.

376. Shanebeck KD, Maliszewski CR, Kennedy MK et al. Regulation of murine B cell growth and differentiation by CD30 ligand. Eur J Immunol 1995; 25:2147-2153.

377. Lindhout E, Lakeman A, Mevissen ML et al. Functionally active Epstein-Barr virus-transformed follicular dendritic cell-like cell lines. J Exp Med 1994; 179:1173-1184.

378. Clark EA, Grabstein KH, Gown AM et al. Activation of B lymphocyte maturation by a human follicular dendritic cell line, FDC-1 155. J Immunol 1995; 155:545-555.

379. Tsunoda R, Bosseloir A, Onozaki K et al. Human follicular dendritic cells in vitro and follicular dendritic-cell-like cells. Cell Tissue Res 1997; 288:381-389.

380. MacLennan IC, Gulbranson-Judge A, Toellner KM et al. The changing preference of T and B cells for partners as T-dependent antibody responses develop. Immunol Rev 1997; 156:53-66.

381. Kosco MH, Pflugfelder E, Gray D. Follicular dendritic cell-dependent adhesion and proliferation of B cells in vitro. J Immunol 1992; 148:2331-2339.

382. Wu J, Qin D, Burton GF et al. Follicular dendritic cell-derived antigen and accessory activity in initiation of memory IgG responses in vitro. J Immunol 1996; 157:3404-3411.

383. Burton GF, Kupp LI, McNalley EC et al. Follicular dendritic cells and B cell chemotaxis. Eur J Immunol 1995; 25:1105-1108.

384. Kosco-Vilbois MH, Gray D, Scheidegger D et al. Follicular dendritic cells help resting B cells to become effective antigen-presenting cells: Induction of B7/BB1 and upregulation of major histocompatibility complex class II molecules. J Exp Med 1993; 178:2055-2066.

385. Reynaud CA, Anquez V, Grimal H et al. A hyperconversion mechanism generates the chicken light chain preimmune repertoire. Cell 1987; 48:379-388.
386. Reynaud CA, Dahan A, Anquez V et al. Somatic hyperconversion diversifies the single Vh gene of the chicken with a high incidence in the D region. Cell 1989; 59:171-183.
387. Parvari R, Ziv E, Lantner F et al. Somatic diversification of chicken immunoglobulin light chains by point mutations. Proc Natl Acad Sci USA 1990; 87:3072-3076.
388. Voss EW, Jr., Watt RM. Comparison of the microenvironment of chicken and rabbit antibody active sites. Adv Exp Med Biol 1977; 88:391-401.
389. Arakawa H, Furusawa S, Ekino S et al. Immunoglobulin gene hyperconversion ongoing in chicken splenic germinal centers. Embo J 1996; 15:2540-2546.
390. David V, Folk NL, Maizels N. Germ line variable regions that match hypermutated sequences in genes encoding murine anti-hapten antibodies. Genetics 1992; 132:799-811.
391. Ford JE, McHeyzer-Williams MG, Lieber MR. Analysis of individual immunoglobulin lambda light chain genes amplified from single cells is inconsistent with variable region gene conversion in germinal-center B cell somatic mutation. Eur J Immunol 1994; 24:1816-1822.
392. Rogerson BJ. Somatic hypermutation of VHS107 genes is not associated with gene conversion among family members. Int Immunol 1995; 7:1225-1235.
393. Parng CL, Hansal S, Goldsby RA et al. Gene conversion contributes to Ig light chain diversity in cattle. J Immunol 1996; 157:5478-5486.
394. Reynaud CA, Mackay CR, Muller RG et al. Somatic generation of diversity in a mammalian primary lymphoid organ: The sheep ileal Peyer's patches. Cell 1991; 64:995-1005.
395. Siskind GW, Dunn P, Walker JG. Studies on the control of antibody synthesis. II. Effect of antigen dose and of suppression by passive antibody on the affinity of antibody synthesized. J Exp Med 1968; 127:55-66.
396. Weigert MG, Cesari IM, Yonkovich SJ et al. Variability in the lambda light chain sequences of mouse antibody. Nature 1970; 228:1045-1047.
397. Winter DB, Gearhart PJ. Somatic hypermutation: Another piece in the hypermutation puzzle. Curr Biol 1995; 5:1345-1346.
398. Jolly CJ, Wagner SD, Rada C et al. The targeting of somatic hypermutation. Semin Immunol 1996; 8:159-168.
399. Betz AG, Milstein C, Gonzalez-Fernandez A et al. Elements regulating somatic hypermutation of an immunoglobulin kappa gene: Critical role for the intron enhancer/matrix attachment region. Cell 1994; 77:239-248.
400. Peters A, Storb U. Somatic hypermutation of immunoglobulin genes is linked to transcription initiation. Immunity 1996; 4:57-65.

401. Berek C, Milstein C. Mutation drift and repertoire shift in the maturation of the immune response. Immunol Rev 1987; 96:23-41.

402. Berek C, Apel M. Somatic changes in the immune response to the hapten 2-phenyloxazolone. In Progr in Immunology Melchers et al Eds Springer-Verlag Berlin 1989; 7:99-105.

403. Berek C, Berger A, Apel M. Maturation of the immune response in germinal centers. Cell 1991; 67:1121-1129.

404. Ziegner M, Berek C. Analysis of germinal centres in the immune response to oxazolone. Adv Exp Med Biol 1994; 355:201-205.

405. Weiss U, Rajewsky K. The repertoire of somatic antibody mutants accumulating in the memory compartment after primary immunization is restricted through affinity maturation and mirrors that expressed in the secondary response. J Exp Med 1990; 172:1681-1689.

406. Cumano A, Rajewsky K. Structure of primary anti-(4-hydroxy-3-nitro-phenyl)acetyl (NP) antibodies in normal and idiotypically suppressed C57BL/6 mice. Eur J Immunol 1985; 15:512-520.

407. Makela O, Karjalainen K. Inherited immunoglobulin idiotypes of the mouse. Immunol Rev 1977; 34:119-138.

408. Claflin JL, Berry J, Flaherty D et al. Somatic evolution of diversity among anti-phosphocholine antibodies induced with Proteus morganii. J Immunol 1987; 138:3060-3068.

409. MacLennan IC, Gray D. Antigen-driven selection of virgin and memory B cells. Immunol Rev 1986; 91:61-85.

410. Dunn-Walters DK, Isaacson PG, Spencer J. Analysis of mutations in immunoglobulin heavy chain variable region genes of microdissected marginal zone (MGZ) B cells suggests that the MGZ of human spleen is a reservoir of memory B cells. J Exp Med 1995; 182:559-566.

411. Kuppers R, Zhao M, Hansmann ML et al. Tracing B cell development in human germinal centres by molecular analysis of single cells picked from histological sections. Embo J 1993; 12:4955-4967.

412. Berek C. The development of B cells and the B-cell repertoire in the micro-environment of the germinal center. Immunol Rev 1992; 126:5-19.

413. Leanderson T, Kallberg E, Gray D. Expansion, selection and mutation of antigen-specific B cells in germinal centers. Immunol Rev 1992; 126:47-61.

414. Chu YW, Marin E, Fuleihan R et al. Somatic mutation of human immunoglobulin V genes in the X-linked HyperIgM syndrome. J Clin Invest 1995; 95:1389-1393.

415. Lebecque S, de Bouteiller O, Arpin C et al. Germinal center founder cells display propensity for apoptosis before onset of somatic mutation. J Exp Med 1997; 185:563-571.

416. Jacob J, Przylepa J, Miller C et al. In situ studies of the primary immune response to (4-hydroxy-3-nitrophenyl)acetyl. III. The kinetics of V region mutation and selection in germinal center B cells. J Exp Med 1993; 178:1293-1307.

417. Ziegner M, Steinhauser G, Berek C. Development of antibody diversity in single germinal centers: Selective expansion of high-affinity variants. Eur J Immunol 1994; 24:2393-2400.

418. Kelsoe G. B cell diversification and differentiation in the periphery. J Exp Med 1994; 180:5-6.

419. Kallberg E, Gray D, Leanderson T. The effect of carrier and carrier priming on the kinetics and pattern of somatic mutation in the V chi Ox1 gene. Eur J Immunol 1995; 25:2349-2354.

420. Sordat B, Sordat M, Hess MW et al. Specific antibody within lymphoid germinal center cells of mice after primary immunization with horseradish peroxidase: A light and electron microscopic study. J Exp Med 1970; 131:77-91.

421. Ridderstad A, Nossal GJ, Tarlinton DM. The xid mutation diminishes memory B cell generation but does not affect somatic hypermutation and selection. J Immunol 1996; 157:3357-3365.

422. McKean D, Huppi K, Bell M et al. Generation of antibody diversity in the immune response of BALB/c mice to influenza virus hemagglutinin. Proc Natl Acad Sci USA 1984; 81:3180-3184.

423. Siekevitz M, Kocks C, Rajewsky K et al. Analysis of somatic mutation and class switching in naive and memory B cells generating adoptive primary and secondary responses. Cell 1987; 48:757-770.

424. Shlomchik MJ, Marshak-Rothstein A, Wolfowicz CB et al. The role of clonal selection and somatic mutation in autoimmunity. Nature 1987; 328:805-811.

425. Decker DJ, Linton P-J, Zaharevitz S et al. Defining subsets of naive and memory B cells based on the ability of their progeny to somatically mutate in vitro. Immunity 1995; 2:195-203.

426. Shan H, Shlomchik M, Weigert M. Heavy-chain class switch does not terminate somatic mutation. J Exp Med 1990; 172:531-536.

427. Wu H, Pelkonen E, Knuutila S et al. A human follicular lymphoma B cell line hypermutates its functional immunoglobulin genes in vitro. Eur J Immunol 1995; 25:3263-3269.

427a. Denepoux S, Razanajaona D, Blanchard D et al. Induction of somatic mutation in a human B cell line in vitro. Immunity 1997; 6:35-46.

428. Kallberg E, Jainandunsing S, Gray D et al. Somatic mutation of immunoglobulin V genes in vitro. Science 1996; 271:1285-1289.

429. Du Pasquier L. Phylogeny of B-cell development. Curr Opin Immunol 1993; 5:185-193.

430. Cohen N. Phylogenetic emergence of lymphoid tissues and cells. In: Marchalonis JJ, ed. The Lymphocyte: Structure and Function Part 1. New York: Marcel Dekker, 1977:149-202.

431. Nahm MH, Kroese FG, Hoffmann JW. The evolution of immune memory and germinal centers. Immunol Today 1992; 13:438-441.

432. Wilson M, Hsu E, Marcuz A et al. What limits affinity maturation of antibodies in Xenopus—the rate of somatic mutation or the ability to select mutants? Embo J 1992; 11:4337-4347.

433. Motyka B, Griebel PJ, Reynolds JD. Agents that activate protein kinase C rescue sheep ileal Peyer's patch B cells from apoptosis. Eur J Immunol 1993; 23:1314-1321.
434. Asakawa J, Tsiagbe VK, Thorbecke GJ. Protection against apoptosis in chicken bursa cells by phorbol ester in vitro. Cell Immunol 1993; 147:180-187.
435. Osmond DG. Proliferation kinetics and the lifespan of B cells in central and peripheral lymphoid organs. Curr Opin Immunol 1991; 3:179-185.
436. Osmond DG. Production and selection of B lymphocytes in bone marrow: Lymphostromal interactions and apoptosis in normal, mutant and transgenic mice. Adv Exp Med Biol 1994; 355:15-20.
437. Nossal GJ. B lymphocyte physiology: The beginning and the end. Ciba Found Symp 1997; 204:220-231.
438. Smith KG, Weiss U, Rajewsky K et al. Bcl-2 increases memory B cell recruitment but does not perturb selection in germinal centers. Immunity 1994; 1:803-813.
439. Papavasiliou F, Casellas R, Suh H et al. V(D)J recombination in mature B cells: A mechanism for altering antibody responses. Science 1997; 278:298-301.
440. Han S, Dillon SR, Zheng B et al. V(D)J recombinase activity in a subset of germinal center B lymphocytes. Science 1997; 278:301-305.
441. Linton PJ, Rudie A, Klinman NR. Tolerance susceptibility of newly generating memory B cells. J Immunol 1991; 146:4099-4104.
442. Linton PJ, Klinman NR. The use of the splenic focus assay to study B cell tolerance. Immunometh 1993; 2:95-103.
443. Klinman NR. The "clonal selection hypothesis" and current concepts of B cell tolerance. Immunity 1996; 5:189-195.
444. Radic MZ, Erikson J, Litwin S et al. B lymphocytes may escape tolerance by revising their antigen receptors. J Exp Med 1993; 177:1165-1173.
445. Gay D, Saunders T, Camper S et al. Receptor editing: An approach by autoreactive B cells to escape tolerance. J Exp Med 1993; 177:999-1008.
446. Tiegs SL, Russell DM, Nemazee D. Receptor editing in self-reactive bone marrow B cells. J Exp Med 1993; 177:1009-1020.
447. Hartley SB, Goodnow CC. Censoring of self-reactive B cells with a range of receptor affinities in transgenic mice expressing heavy chains for a lysozyme-specific antibody. Int Immunol 1994; 6:1417-1425.
448. Shokat KM, Goodnow CC. Antigen-induced B-cell death and elimination during germinal-centre immune responses. Nature 1995; 375:334-338.
449. Pulendran B, Kannourakis G, Nouri S et al. Soluble antigen can cause enhanced apoptosis of germinal-centre B cells. Nature 1995; 375:331-334.
450. Hahne M, Renno T, Schroeter M et al. Activated B cells express functional Fas ligand. Eur J Immunol 1996; 26:721-724.

451. Tighe H, Warnatz K, Brinson D et al. Peripheral deletion of rheumatoid factor B cells after abortive activation by IgG. Proc Natl Acad Sci USA 1997; 94:646-651.
452. Secord EA, Edington JM, Thorbecke GJ. The Emu-bcl-2 transgene enhances antigen-induced germinal center formation in both BALB/c and SJL mice but causes age-dependent germinal center hyperplasia only in the lymphoma-prone SJL strain. Am J Pathol 1995; 147:422-433.
453. Strasser A, Whittingham S, Vaux DL et al. Enforced BCL2 expression in B-lymphoid cells prolongs antibody responses and elicits autoimmune disease. Proc Natl Acad Sci USA 1991; 88:8661-8665.
454. Szakal AK, Taylor JK, Smith JP et al. Kinetics of germinal center development in lymph nodes of young and aging immune mice. Anat Rec 1990; 227:475-485.
455. Hanna MG, Jr., Nettesheim P, Ogden L et al. Reduced immune potential of aged mice: Significance of morphologic changes in lymphatic tissue. Proc Soc Exp Biol Med 1967; 125:882-887.
456. Kraft R, Bachmann M, Bachmann K et al. Satisfactory primary tetanus antitoxin responses but markedly reduced germinal centre formation in first draining lymph nodes of ageing mice. Clin Exp Immunol 1987; 67:447-453.
457. Legge JS, Austin CM. Antigen localization and the immune response as a function of age. Aust J Exp Biol Med Sci 1968; 46:361-365.
458. Coico RF, Thorbecke GJ. Role of germinal centers in the generation of B cell memory. Folia Microbiol (Praha) 1985; 30:196-202.
459. Miller C, Kelsoe G. Ig VH hypermutation is absent in the germinal centers of aged mice. J Immunol 1995; 155:3377-3384.
460. Linton PJ. The status of progenitors of memory B cells in aged mice. AGING: Immunol Infect Dis 1993; 4:35-46.
461. Szakal AK, Taylor JK, Smith JP et al. Morphometry and kinetics of antigen transport and developing antigen-retaining reticulum of follicular dendritic cells in lymph nodes of aging immune mice. AGING: Immunol Infect Dis 1988; 1:7-22.
462. Burton GF, Kosco MH, Szakal AK et al. Iccosomes and the secondary antibody response. Immunology 1991; 73:271-276.
463. Kapasi ZF, Tew JG, Szakal AK. Germinal center development and the secondary antibody response following bone marrow and thymus transplantation in old mice. AGING: Immunol Infect Dis 1993; 4:77-94.
464. Klaus GG, Kunkl A. The role of T cells in B cell priming and germinal centre development. Adv Exp Med Biol 1982; 149:743-749.
465. Gonzalez-Fernandez A, Gilmore D, Milstein C. Age-related decrease in the proportion of germinal center B cells from mouse Peyer's patches is accompanied by an accumulation of somatic mutations in their immunoglobulin genes. Eur J Immunol 1994; 24:2918-2921.
466. Smith JP, Lister AM, Tew JG et al. Kinetics of the tingible body macrophage response in mouse germinal center development and its depression with age. Anat Rec 1991; 229:511-520.

467. Rosenberg YJ, Kosco MH, Lewis MG et al. Changes in follicular dendritic cell and CD8⁺ cell function in macaque lymph nodes following infection with SIV251. Adv Exp Med Biol 1993; 329:417-423.

468. Spiegel H, Herbst H, Niedobitek G et al. Follicular dendritic cells are a major reservoir for human immunodeficiency virus type 1 in lymphoid tissues facilitating infection of CD4⁺ T-helper cells. Am J Pathol 1992; 140:15-22.

469. Devergne O, Peuchmaur M, Crevon MC et al. Activation of cytotoxic cells in hyperplastic lymph nodes from HIV-infected patients. AIDS 1991; 5: 1071-1079.

470. Tenner-Racz K, Racz P, Thome C et al. Cytotoxic effector cell granules recognized by the monoclonal antibody TIA-1 are present in CD8⁺ lymphocytes in lymph nodes of human immunodeficiency virus-1-infected patients. Am J Pathol 1993; 142:1750-1758.

471. Tenner-Racz K. Human immunodeficiency virus associated changes in germinal centers of lymph nodes and relevance to impaired B-cell function. Lymphology 1988; 21:36-43.

472. Tenner-Racz K, Racz P, Schmidt H et al. Immunohistochemical, electron microscopic and in situ hybridization evidence for the involvement of lymphatics in the spread of HIV-1. AIDS 1988; 2:299-309.

473. Laman JD, Claassen E, Van Rooijen N et al. Immune complexes on follicular dendritic cells as a target for cytolytic cells in AIDS. AIDS 1989; 3:543-544.

474. Koopman G, Wever PC, Ramkema MD et al. Expression of granzyme B by cytotoxic T lymphocytes in the lymph nodes of HIV-infected patients. AIDS Res Hum Retroviruses 1997; 13:227-233.

475. Philpott KL, Viney JL, Kay G et al. Lymphoid development in mice congenitally lacking T cell receptor alpha beta-expressing cells. Science 1992; 256:1448-1452.

476. Dianda L, Gulbranson-Judge A, Pao W et al. Germinal center formation in mice lacking alpha beta T cells. Eur J Immunol 1996; 26:1603-1607.

477. Fehling HJ, Viville S, van Ewijk W et al. Fine-tuning of MHC class II gene expression in defined microenvironments. Trends Genet 1989; 5:342-347.

478. Fischer MB, Ma M, Goerg S et al. Regulation of the B cell response to T-dependent antigens by classical pathway complement. J Immunol 1996; 157:549-556.

479. Romball CG, Ulevitch RJ, Weigle WO. Role of C3 in the regulation of a splenic PFC response in rabbits. J Immunol 1980; 124:151-155.

480. Klaus GG, Humphrey JH. The generation of memory cells. I. The role of C3 in the generation of B memory cells. Immunology 1977; 33:31-40.

481. Van den Berg TK, Dopp EA, Daha MR et al. Selective inhibition of immune complex trapping by follicular dendritic cells with monoclonal antibodies against rat C3. Eur J Immunol 1992; 22:957-962.

482. Ahearn JM, Fischer MB, Croix D et al. Disruption of the Cr2 locus results in a reduction in B-1a cells and in an impaired B cell response to T-dependent antigen. Immunity 1996; 4:251-262.

483. Croix DA, Ahearn JM, Rosengard AM et al. Antibody response to a T-dependent antigen requires B cell expression of complement receptors. J Exp Med 1996; 183:1857-1864.
484. Carroll MC, Fischer MB. Complement and the immune response. Curr Opin Immunol 1997; 9:64-69.
485. Maecker HT, Levy S. Normal lymphocyte development but delayed humoral immune response in CD81-null mice. J Exp Med 1997; 185:1505-1510.
486. Matsumoto M, Fukuda W, Circolo A et al. Abrogation of the alternative complement pathway by targeted deletion of murine factor B. Proc Natl Acad Sci USA 1997; 94:8720-8725.
487. De Togni P, Goellner J, Ruddle NH et al. Abnormal development of peripheral lymphoid organs in mice deficient in lymphotoxin. Science 1994; 264:703-707.
488. Banks TA, Rouse BT, Kerley MK et al. Lymphotoxin-alpha-deficient mice. Effects on secondary lymphoid organ development and humoral immune responsiveness. J Immunol 1995; 155:1685-1693.
489. Mariathasan S, Matsumoto M, Baranyay F et al. Absence of lymph nodes in lymphotoxin-alpha(LT alpha)-deficient mice is due to abnormal organ development, not defective lymphocyte migration. J Inflamm 1995; 45:72-78.
489a. Gonzalez M, Mackay F, Browning JL et al. The sequential role of lymphotoxin and B cells in the development of splenic follicles. J Exp Med 1998; 187:997-1007.
490. Sacca R, Turley S, Soong L et al. Transgenic expression of lymphotoxin restores lymph nodes to lymphotoxin-α-deficient mice. J Immunol 1997; 159:4252-4260.
491. Liu YJ, Banchereau J. Mutant mice without B lymphocyte follicles. J Exp Med 1996; 184:1207-1211.
492. Matsumoto M, Mariathasan S, Nahm MH et al. Role of lymphotoxin and the type I TNF receptor in the formation of germinal centers. Science 1996; 271:1289-1291.
493. Le Hir M, Bluethmann H, Kosco-Vilbois MH et al. Differentiation of follicular dendritic cells and full antibody responses require tumor necrosis factor receptor-1 signaling. J Exp Med 1996; 183:2367-2372.
494. Koni PA, Sacca R, Lawton P et al. Distinct roles in lymphoid organogenesis for lymphotoxins alpha and beta revealed in lymphotoxin beta-deficient mice. Immunity 1997; 6:491-500.
495. Neumann B, Luz A, Pfeffer K et al. Defective Peyer's patch organogenesis in mice lacking the 55-kD receptor for tumor necrosis factor. J Exp Med 1996; 184:259-264.
496. Pasparakis M, Alexopoulou L, Grell M et al. Peyer's patch organogenesis is intact yet formation of B lymphocyte follicles is defective in peripheral lymphoid organs of mice deficient for tumor necrosis factor and its 55-kDa receptor. Proc Natl Acad Sci USA 1997; 94:6319-6323.
497. Alimzhanov MB, Kuprash DV, Kosco-Vilbois MH et al. Abnormal development of secondary lymphoid tissues in lymphotoxin β-deficient mice. Immunology 1997; 94:9302-9307.

498. Alimzhanov MB, Kuprash DV, Kosco-Vilbois MH et al. Abnormal development of secondary lymphoid tissues in lymphotoxin beta-deficient mice. Proc Natl Acad Sci USA 1997; 94:9302-9307.
499. Mackay F, Majeau GR, Lawton P et al. Lymphotoxin but not tumor necrosis factor functions to maintain splenic architecture and humoral responsiveness in adult mice. Eur J Immunol 1997; 27: 2033-2042.
500. Fu YX, Molina H, Matsumoto M et al. Lymphotoxin-alpha (LTalpha) supports development of splenic follicular structure that is required for IgG responses. J Exp Med 1997; 185:2111-2120.
501. Rennert PD, Browning JL, Mebius R et al. Surface lymphotoxin alpha/beta complex is required for the development of peripheral lymphoid organs. J Exp Med 1996; 184:1999-2006.
502. Matsumoto M, Fu YX, Molina H et al. Lymphotoxin-alpha-deficient and TNF receptor-I-deficient mice define developmental and functional characteristics of germinal centers. Immunol Rev 1997; 156:137-144.
503. Ettinger R, Browning JL, Michie SA et al. Disrupted splenic architecture, but normal lymph node development in mice expressing a soluble lymphotoxin-beta receptor-IgG1 fusion protein. Proc Natl Acad Sci USA 1996; 93:13102-13107.
504. Kratz A, Campos-Neto A, Hanson MS et al. Chronic inflammation caused by lymphotoxin is lymphoid neogenesis. J Exp Med 1996; 183:1461-1472.
505. Forster R, Wolf I, Kaiser E et al. Selective expression of the murine homologue of the G-protein-coupled receptor BLR1 in B cell differentiation, B cell neoplasia and defined areas of the cerebellum. Cell Mol Biol 1994; 40:381-387.
505a. Legler, DF, Loetscher M, Roos RS et al. B cell-attracting chemokine 1, a human CXC chemokine expressed in lymphoid tissues, selectively attracts B lymphoctyes via BLR1/CXCR5. J Exp Med 1998; 187:655-600.
506. Dobner T, Wolf I, Emrich T et al. Differentiation-specific expression of a novel G protein-coupled receptor from Burkitt's lymphoma. Eur J Immunol 1992; 22:2795-2799.
507. Burgstahler R, Kempkes B, Steube K et al. Expression of the chemokine receptor BLR2/EBI1 is specifically transactivated by Epstein-Barr virus nuclear antigen 2. Biochem Biophys Res Commun 1995; 215:737-743.
508. Nagira M, Imai T, Hieshima K et al. Molecular cloning of a novel human CC chemokine secondary lymphoid-tissue chemokine that is a potent chemoattractant for lymphocytes and mapped to chromosome 9p13. J Biol Chem 1997; 272:19518-19524.
509. Yoshida R, Imai T, Hieshima K et al. Molecular cloning of a novel human CC chemokine EBI1-ligand chemokine that is a specific functional ligand for EBI1, CCR7. J Biol Chem 1997; 272:13803-13809.

509a. Turner M, Gulbranson-Judge A, Quinn ME et al. Syk tyrosine kinase is required for the positive selection of immature B cells into the recirculating B cell pool. J Exp Med 1997; 186:2013-2021.

510. Miyawaki S, Nakamura Y, Suzuka H et al. A new mutation, aly, that induces a generalized lack of lymph nodes accompanied by immunodeficiency in mice. Eur J Immunol 1994; 24:429-434.

511. Shinkura R, Matsuda F, Sakiyama T et al. Defects of somatic hypermutation and class switching in alymphoplasia (aly) mutant mice. Int Immunol 1996; 8:1067-1075.

512. Caamano J, Rizzo C, Druham SK et al. NF-κB2 (p100/p52) is required for normal splenic microarchitecture and B cell-mediated immune responses. J Exp Med 1998; 187:185-196.

513. Franzoso G, Carlson L, Scharton-Kersten T et al. Critical roles for the Bcl-3 oncoprotein in T cell-mediated immunity, splenic microarchitecture, and germinal center reactions. Immunity 1997; 6:479-490.

514. Carrasco D, Weih F, Bravo R. Developmental expression of the mouse c-rel proto-oncogene in hematopoietic organs. Development 1994; 120:2991-3004.

515. Kontgen F, Grumont RJ, Strasser A et al. Mice lacking the c-rel proto-oncogene exhibit defects in lymphocyte proliferation, humoral immunity, and interleukin-2 expression. Genes Dev 1995; 9:1965-1977.

516. Gerondakis S, Strasser A, Metcalf D et al. Rel-deficient T cells exhibit defects in production of interleukin 3 and granulocyte-macrophage colony-stimulating factor. Proc Natl Acad Sci USA 1996; 93:3405-3409.

517. Weih F, Carrasco D, Durham SK et al. Multiorgan inflammation and hematopoietic abnormalities in mice with a targeted disruption of RelB, a member of the NF-kappa B/Rel family. Cell 1995; 80:331-340.

518. Naspetti M, Aurrand-Lions M, DeKoning J et al. Thymocytes and RelB-dependent medullary epithelial cells provide growth-promoting and organization signals, respectively, to thymic medullary stromal cells. Eur J Immunol 1997; 27:1392-1397.

519. Weih F, Durham SK, Barton DS et al. Both multiorgan inflammation and myeloid hyperplasia in RelB-deficient mice are T cell dependent. J Immunol 1996; 157:3974-3979.

520. Weih F, Durham SK, Barton DS et al. p50-Nf-kappaB complexes partially compensate for the absence of RelB: Severely increased pathology in p50(-/-)relB(-/-) double-knockout mice. J Exp Med 1997; 185:1359-1370.

521. Doi TS, Takahashi T, Taguchi O et al. NF-kappa B RelA-deficient lymphocytes: Normal development of T cells and B cells, impaired production of IgA and IgG1 and reduced proliferative responses. J Exp Med 1997; 185:953-961.

522. Horwitz BH, Scott ML, Cherry SR et al. Failure of lymphopoiesis after adoptive transfer of NF-kappaB-deficient fetal liver cells. Immunity 1997; 6:765-772.

523. Beg AA, Sha WC, Bronson RT et al. Constitutive NF-kappa B activation, enhanced granulopoiesis, and neonatal lethality in I kappa B alpha-deficient mice. Genes Dev 1995; 9:2736-2746.
524. Fukuda T, Yoshida T, Okada S et al. Disruption of the Bcl6 gene results in an impaired germinal center formation. J Exp Med 1997; 186:439-448.
525. Ye BH, Cattoretti G, Shen Q et al. The Bcl-6 proto-oncogene controls germinal-centre formation and Th2-type inflammation. Nat Genet 1997; 16:161-170.
526. Dent AL, Shaffer AL, Yu X et al. Control of inflammation, cytokine expression, and germinal center formation by Bcl-6. Science 1997; 276:589-592.
527. Nishizumi H, Taniuchi I, Yamanashi Y et al. Impaired proliferation of peripheral B cells and indication of autoimmune disease in lyn-deficient mice. Immunity 1995; 3:549-560.
528. Wang JH, Nichogiannopoulou A, Wu L et al. Selective defects in the development of the fetal and adult lymphoid system in mice with an Ikaros null mutation. Immunity 1996; 5:537-549.
529. Hodge MR, Ranger AM, Charles de la Brousse F et al. Hyperproliferation and dysregulation of IL-4 expression in NF-ATp-deficient mice. Immunity 1996; 4:397-405.
530. Yasue T, Nishizumi H, Aizawa S et al. A critical role of lyn and fyn for B cell responses to CD38 ligation and interleukin 5. Proc Natl Acad Sci USA 1997; 94:10307-10312.
531. Worm M, Geha RS. CD40 ligation induces lymphotoxin alpha gene expression in human B cells. Int Immunol 1994; 6:1883-1890.
532. Worm M, Geha RS. CD40-mediated lymphotoxin alpha expression in human B cells is tyrosine kinase dependent. Eur J Immunol 1995; 25:2438-2444.
533. Hanissian SH, Geha RS. Jak3 is associated with CD40 and is critical for CD40 induction of gene expression in B cells. Immunity 1997; 6:379-387.
534. Neumann M, Wohlleben G, Chuvpilo S et al. CD40, but not lipopolysaccharide and anti-IgM stimulation of primary B lymphocytes, leads to a persistent nuclear accumulation of RelB. J Immunol 1996; 157:4862-4869.
535. Borriello F, Sethna MP, Boyd SD et al. B7-1 and B7-2 have overlapping, critical roles in immunoglobulin class switching and germinal center formation. Immunity 1997; 6:303-313.
536. Hough MR, Chappel MS, Sauvageau G et al. Reduction of early B lymphocyte precursors in transgenic mice overexpressing the murine heat-stable antigen. J Immunol 1996; 156:479-488.
537. Siegal A, Kopel S, Leibovici J. Histological changes in spleen and lymph nodes of mice administered cyclophosphamide and levan. Cell Tissue Res 1986; 245:183-188.
538. Thorbecke G, Lerman SP. Germinal centers and their role in immune responses. Adv Exp Med Biol 1976; 73 PT-A:83-100.

539. Durkin HG, Thorbecke GJ. Relationship of germinal centers in lymphoid tissue to immunologic memory. V. The effect of prednisolone administered after the peak of the primary response. J Immunol 1971; 106:1079-1085.

540. Antoine N, Daukandt M, Heinen E et al. In vitro and in vivo stimulation of the murine immune system by AGM-1470, a potent angiogenesis inhibitor. Am J Pathol 1996; 148:393-398.

541. Smiley JD, Moore SE, Jr. Molecular mechanisms of autoimmunity. Am J Med Sci 1988; 295:478-496.

542. Hurtenbach U, Gleichmann H, Nagata N et al. Immunity to D-penicillamine: Genetic, cellular, and chemical requirements for induction of popliteal lymph node enlargement in the mouse. J Immunol 1987; 139:411-416.

543. Jacobson BA, Panka DJ, Nguyen KA et al. Anatomy of autoantibody production: Dominant localization of antibody-producing cells to T cell zones in Fas-deficient mice. Immunity 1995; 3:509-519.

544. Khan WN, Nilsson A, Mizoguchi E et al. Impaired B cell maturation in mice lacking Bruton's tyrosine kinase (Btk) and CD40. Int Immunol 1997; 9:395-405.

545. Castigli E, Alt FW, Davidson L et al. CD40-deficient mice generated by recombination-activating gene-2-deficient blastocyst complementation. Proc Natl Acad Sci USA 1994; 91:12135-12139.

546. Rickert RC, Rajewsky K, Roes J. Impairment of T-cell-dependent B-cell responses and B-1 cell development in CD19-deficient mice. Nature 1995; 376:352-355.

547. Engel P, Zhou LJ, Ord DC et al. Abnormal B lymphocyte development, activation, and differentiation in mice that lack or overexpress the CD19 signal transduction molecule. Immunity 1995; 3:39-50.

548. Tsitsikov EN, Gutierrez-Ramos JC, Geha RS. Impaired CD19 expression and signaling, enhanced antibody response to type II T independent antigen and reduction of B-1 cells in CD81-deficient mice. Proc Natl Acad Sci USA 1997; 94:10844-10849.

549. Sato S, Miller AS, Inaoki M et al. CD22 is both a positive and negative regulator of B lymphocyte antigen receptor signal transduction: Altered signaling in CD22-deficient mice. Immunity 1996; 5:551-562.

550. Otipoby KL, Andersson KB, Draves KE et al. CD22 regulates thymus-independent responses and the lifespan of B cells. Nature 1996; 384:634-637.

551. O'Keefe T, Williams GT, Davies SL et al. Hyperresponsive B cells in CD22-deficient mice. Science 1996; 274:798-801.

552. Renshaw BR, Fanslow WC III, Armitage RJ et al. Humoral immune responses in CD40 ligand-deficient mice. J Exp Med 1994; 180:1889-1900.

553. Ostberg JR, Dragone LL, Driskell T et al. Disregulated expression of CD43 (leukosialin, sialophorin) in the B cell lineage leads to immunodeficiency. J Immunol 1996; 157:4876-4884.

554. Ostberg J, Dragone L, Borello M et al. Expression of mouse CD43 in the B cell lineage of transgenic mice causes impaired immune responses to T-independent antigens. Eur J Immunol 1997; 27:2152-2159.

555. Ferguson SE, Han S, Kelsoe G et al. CD28 is required for germinal center formation. J Immunol 1996; 156:4576-4581.

556. Lane P, Burdet C, Hubele S et al. B cell function in mice transgenic for mCTLA4-H gamma 1: Lack of germinal centers correlated with poor affinity maturation and class switching despite normal priming of CD4⁺ T cells. J Exp Med 1994; 179:819-830.

557. Ronchese F, Hausmann B, Hubele S et al. Mice transgenic for a soluble form of murine CTLA-4 show enhanced expansion of antigen-specific CD4⁺ T cells and defective antibody production in vivo. J Exp Med 1994; 179:809-817.

558. Eichelberger M, McMickle A, Blackman M et al. Functional analysis of the TCR alpha-beta+ cells that accumulate in the pneumonic lung of influenza virus-infected TCR-alpha-/-mice. J Immunol 1995; 154:1569-1576.

559. Mombaerts P, Clarke AR, Rudnicki MA et al. Mutations in T-cell antigen receptor genes alpha and beta block thymocyte development at different. Nature 1992; 360:225-231.

560. Pao W, Wen L, Smith AL et al. Gamma delta T cell help of B cells is induced by repeated parasitic infection, in the absence of other T cells. Curr Biol 1996; 6:1317-1325.

561. Yu P, Kosco-Vilbois M, Richards M et al. Negative feedback regulation of IgE synthesis by murine CD23. Nature 1994; 369:753-756.

562. Takai T, Ono M, Hikida M et al. Augmented humoral and anaphylactic responses in Fc gamma RII-deficient mice. Nature 1996; 379:346-349.

563. Frenette PS, Mayadas TN, Rayburn H et al. Susceptibility to infection and altered hematopoiesis in mice deficient in both P- and E-selectins. Cell 1996; 84:563-574.

564. Steeber DA, Green NE, Sato S et al. Humoral immune responses in L-selectin-deficient mice. J Immunol 1996; 157:4899-4907.

565. Matsumoto M, Lo SF, Carruthers CJ et al. Affinity maturation without germinal centres in lymphotoxin-alpha-deficient mice. Nature 1996; 382:462-466.

566. Sadlack B, Lohler J, Schorle H et al. Generalized autoimmune disease in interleukin-2-deficient mice is triggered by an uncontrolled activation and proliferation of CD4⁺ T cells. Eur J Immunol 1995; 25:3053-3059.

567. Schijns VE, Haagmans BL, Rijke EO et al. IFN-gamma receptor-deficient mice generate antiviral Th1-characteristic cytokine profiles but altered antibody responses. J Immunol 1994; 153:2029-2037.

568. Kamijo R, Le J, Shapiro D et al. Mice that lack the interferon-gamma receptor have profoundly altered responses to infection with Bacillus Calmette-Guerin and subsequent challenge with lipopolysaccharide. J Exp Med 1993; 178:1435-1440.

569. Yoshida T, Ikuta K, Sugaya H et al. Defective B-1 cell development and impaired immunity against Angiostrongylus cantonensis in IL-5R alpha-deficient mice. Immunity 1996; 4:483-494.

570. Nielsen PJ, Georgiev O, Lorenz B et al. B lymphocytes are impaired in mice lacking the transcriptional co-activator Bob1/OCA-B/OBF1. Eur J Immunol 1996; 26:3214-3218.

571. Kim U, Qin XF, Gong S et al. The B-cell-specific transcription coactivator OCA-B/OBF-1/Bob-1 is essential for normal production of immunoglobulin isotypes. Nature 1996; 383:542-547.

572. Schubart DB, Rolink A, Kosco-Vilbois MH et al. B-cell-specific coactivator OBF-1/OCA-B/Bob1 required for immune response and germinal centre formation. Nature 1996; 383:538-542.

573. Knudson CM, Tung KS, Tourtellotte WG et al. Bax-deficient mice with lymphoid hyperplasia and male germ cell death. Science 1995; 270:96-99.

574. Mittrucker HW, Matsuyama T, Grossman A et al. Requirement for the transcription factor LSIRF/IRF4 for mature B and T lymphocyte function. Science 1997; 275:540-543.

575. Akira S, Yoshida K, Tanaka T et al. Targeted disruption of the IL-6 related genes: gp130 and NF-IL-6. Immunol Rev 1995; 148:221-253.

576. Screpanti I, Musiani P, Bellavia D et al. Inactivation of the IL-6 gene prevents development of multicentric Castleman's disease in C/EBP beta-deficient mice. J Exp Med 1996; 184:1561-1566.

577. Snapper CM, Zelazowski P, Rosas FR et al. B cells from p50/NF-kappa B knockout mice have selective defects in proliferation, differentiation, germ-line CH transcription, and Ig class switching. J Immunol 1996; 156:183-191.

578. Grusby MJ. Stat4- and Stat6-deficient mice as models for manipulating T helper cell responses. Biochem Soc Trans 1997; 25:359-360.

579. DeRocco SE, Iozzo R, Ma XP et al. Ectopic expression of A-myb in transgenic mice causes follicular hyperplasia and enhanced B lymphocyte proliferation. Proc Natl Acad Sci USA 1997; 94:3240-3244.

580. Cyster JG, Goodnow CC. Protein tyrosine phosphatase 1C negatively regulates antigen receptor signaling in B lymphocytes and determines thresholds for negative selection. Immunity 1995; 2:13-24.

581. Pani G, Kozlowski M, Cambier JC et al. Identification of the tyrosine phosphatase PTP1C as a B cell antigen receptor-associated protein involved in the regulation of B cell signaling. J Exp Med 1995; 181:2077-2084.

582. Wang J, Koizumi T, Watanabe T. Altered antigen receptor signaling and impaired Fas-mediated apoptosis of B cells in Lyn-deficient mice. J Exp Med 1996; 184:831-838.

583. Chan VW, Meng F, Soriano P et al. Characterization of the B lymphocyte populations in Lyn-deficient mice and the role of Lyn in signal initiation and down-regulation. Immunity 1997; 7:69-81.

584. Oka Y, Rolink AG, Andersson J et al. Profound reduction of mature B cell numbers, reactivities and serum Ig levels in mice which simultaneously carry the XID and CD40 deficiency genes. Int Immunol 1996; 8:1675-1685.

584a. Delibrias CC, Floettmann JE, Rowe M et al. Downregulated expression of SHP-1 in burkitt lymphomas and germinal center B lymphocytes. J Exp Med 1997; 186:1575-1583.

585. Kitamura D, Roes J, Kuhn R et al. A B cell-deficient mouse by targeted disruption of the membrane exon of the immunoglobulin mu chain gene. Nature 1991; 350:423-426.

586. Alt FW, Rathbun G, Oltz E et al. Function and control of recombination-activating gene activity. Ann N Y Acad Sci 1992; 651:277-294.

587. Nakajima H, Shores EW, Noguchi M et al. The common cytokine receptor gamma chain plays an essential role in regulating lymphoid homeostasis. J Exp Med 1997; 185:189-195.

588. Park SY, Saijo K, Takahashi T et al. Developmental defects of lymphoid cells in Jak3 kinase-deficient mice. Immunity 1995; 3:771-782.

589. Thomas DC, Berg LJ. Peripheral expression of Jak3 is required to maintain T lymphocyte function. J Exp Med 1997; 185:197-206.

590. Robb L, Drinkwater CC, Metcalf D et al. Hematopoietic and lung abnormalities in mice with a null mutation of the common beta subunit of the receptors for granulocyte-macrophage colony-stimulating factor and interleukins 3 and 5. Proc Natl Acad Sci USA 1995; 92:9565-9569.

591. Nishinakamura R, Nakayama N, Hirabayashi Y et al. Mice deficient for the IL-3/GM-CSF/IL-5βc receptor exhibit lung pathology and impaired immune response, while beta IL3 receptor-deficient mice are normal. Immunity 1995; 2:211-222.

592. Cambier JC, Newell MK, Justement LB et al. Ia binding ligands and cAMP stimulate nuclear translocation of PKC in B lymphocytes. Nature 1987; 327:629-632.

593. Cambier JC, Morrison DC, Chien MM et al. Modeling of T cell contact-dependent B cell activation. IL-4 and antigen receptor ligation primes quiescent B cells to mobilize calcium in response to Ia cross-linking. J Immunol 1991; 146:2075-2082.

594. Bonnefoy J-Y, Henchoz S, Hardie D et al. A subset of anti-CD21 antibodies promote the rescue of germinal center B cells from apoptosis. Eur J Immunol 1993; 23:969-972.

595. Kozono Y, Duke RC, Schleicher MS et al. Co-ligation of mouse complement receptors 1 and 2 with surface IgM rescues splenic B cells and WEHI-231 cells from anti-surface IgM-induced apoptosis. Eur J Immunol 1995; 25:1013-1017.

596. Holder M, Grafton G, MacDonald I et al. Engagement of CD20 suppresses apoptosis in germinal center B cells. Eur J Immunol 1995; 25:3160-3164.

597. Lens SM, Tesselaar K, den Drijver BF et al. A dual role for both CD40-ligand and TNF-alpha in controlling human B cell death. J Immunol 1996; 156:507-514.

598. Koopman G, Keehnen RM, Lindhout E et al. Germinal center B cells rescued from apoptosis by CD40 ligation or attachment to follicular dendritic cells, but not by engagement of surface immunoglobulin or adhesion receptors, become resistant to CD95-induced apoptosis. Eur J Immunol 1997; 27:1-7.

599. Lecoanet-Henchoz S, Gauchat JF, Aubry JP et al. CD23 regulates monocyte activation through a novel interaction with the adhesion molecules CD11b-CD18 and CD11c-CD18. Immunity 1995; 3:119-125.

600. White LJ, Ozanne BW, Graber P et al. Inhibition of apoptosis in a human pre-B-cell line by CD23 is mediated via a novel receptor. Blood 1997; 90:234-243.

601. Valentine MA, Licciardi KA. Rescue from anti-IgM-induced programmed cell death by the B cell surface proteins CD20 and CD40. Eur J Immunol 1992; 22:3141-3148.

The Germinal Center Reaction: Influence of FDC Function in Normal, Aged, and Retrovirus Infected Hosts

J.G. Tew, D. Qin, J. Wu, G.F. Burton and A.K. Szakal

Introduction

This review focuses on how follicular dendritic cells (FDC) and associated antigens are involved in germinal center reactions and how changes in FDC function alter humoral immune responses. In normal animals, FDC trap antigen in the form of immune complexes and these antigens are presented to B cells, which then process and present the antigen obtained from the FDC to T cells, providing the help needed for germinal center development, production of memory cells, and the production of high affinity antibody. We will review the involvement of FDC in handling the antigen and describe these steps in some detail. In addition, we are aware of a number of these FDC dependent steps that are abnormal in aged and retrovirus infected mice and men. These abnormalities in FDC function will be discussed in the context of what is happening in young adult animals where germinal center reactions and recall antibody responses are normal.

Both aged and retrovirus infected individuals typically have normal to elevated levels of serum antibody, including immunoglobulin G.[1-3] This suggests that T cells are able to provide enough help for B cells to class switch and indicates that B cells have the ability to make immunoglobulin molecules, including molecules associated

The Biology of Germinal Centers in Lymphoid Tissue, edited by G. Jeanette Thorbecke and Vincent K. Tsiagbe. © 1998 Springer-Verlag and R.G. Landes Company.

with immunological memory. However, attempts to immunize aged or retrovirus infected hosts with simple protein antigens frequently result in responses that are markedly depressed relative to those obtained in normal individuals.[4,5] Furthermore, secondary responses are not much improved over primary responses, indicating that immunological memory is diminished.[4] These defects in humoral immunity may be serious because secondary antibody responses are characterized by a long-lasting production of antigen specific IgG that provides protection against certain infectious diseases for long periods of time.[6,7] Thus, the inability to get a good secondary response and sustain these responses may place aged and retrovirus infected individuals at increased risk for a variety of infectious diseases.

We reasoned that a major problem in eliciting and maintaining recall responses with high affinity antibody in the aged and retrovirus infected individuals may be attributable to a problem with FDC function. Evidence will be reviewed here indicating that FDC and their function is abnormal in aged and retrovirus infected individuals. We reason that reestablishment of FDC function may help restore immunological memory and secondary antibody responses with high affinity antibody. Restoring these recall responses could be critical to establishing and maintaining immunity to a variety of infectious agents in aged and retrovirus infected individuals. However, we are also aware that the FDC trap retroviruses and that the viruses retained on the FDC may be highly infectious.[8] This may complicate efforts to restore immune function in retrovirus infected hosts, but it also suggests that it is important to understand FDC and their functions in the regulation of humoral immune responses. For example, a better understanding of FDC may permit us to manipulate the system to allow germinal centers and humoral immunity to develop and yet avoid problems with infectious viruses retained on the FDC.

Antigen Transport, Iccosome Production, Germinal Center Formation, and Long Term Retention of Ag-Ab Complexes in Normal Mice

In previous reviews[9-12] we presented morphological, kinetic, and functional evidence supporting an antigen pathway leading from the formation of Ag-Ab complexes in a site of infection or immunization to germinal center reactions in the follicles of secondary lymphoid organs. This pathway is an alternative to trapping, endocytosis, and processing of antigen by macrophages and lymphoid dendritic cells. We refer to the transport of antigen along this pathway as the alternative antigen pathway to distinguish it from the macrophage pathway.[9,10] In this section we will highlight the importance of antigen transport in the development of the microenvironment where FDC, B cell and T cell interactions lead to the initiation of the germinal center reaction and the regulation of humoral immunity. Antigen in this pathway results in the induction of germinal centers where B cells proliferate, somatically mutate, begin producing Ab of high affinity leading to affinity maturation, and generate B memory cells.[9,10,13] To help visualize the critical events, a working model of the alternative antigen transport pathway is illustrated in Figure 2.1.

FDC are thought to play a pivotal role in the initiation and maintenance of secondary Ab responses and the microenvironment in which they function is important. Histologically, this microenvironment is the lymphoid follicle and follicles represent a functional building block in secondary lymphoid tissues. Upon appropriate antigenic stimulation, germinal centers develop in lymphoid follicles and germinal centers are the primary sites of B memory cell production and a source of antibody forming cells (AFC).[13-16] This germinal center reaction is dependent on the transport of antigen to the follicle where the interactions between Ag, B cells, and T cells are facilitated by the framework provided by the FDC. These FDC function not as individual units but as integral members of a three-dimensional network "sponge-like" microenvironment (FDC-reticulum) which is formed by the interdigitation of dendritic processes from neighboring FDC.

Since FDC in these reticula have the capacity to retain antigen on their surfaces (in the form of immune complexes) for extended periods (i.e., months to years),[17] it was originally referred to as the antigen retaining reticulum.[18] This microenvironment, located in the light zone of secondary follicles, is significant in that it brings together the FDC-retained Ag, B cells, and T helper cells for the cellular interactions needed for the development of germinal centers.[9] The functional role of the follicular microenvironment in the germinal center reaction is illustrated by the cellular events leading to an early phase of AFC production and a late phase of B memory cell generation.[13,14]

The alternative antigen transport pathway is an adaptation that uses small amounts of antigen efficiently in an environment rich in specific Ab. This may relate to the fact that Ag-Ab complexes on FDC

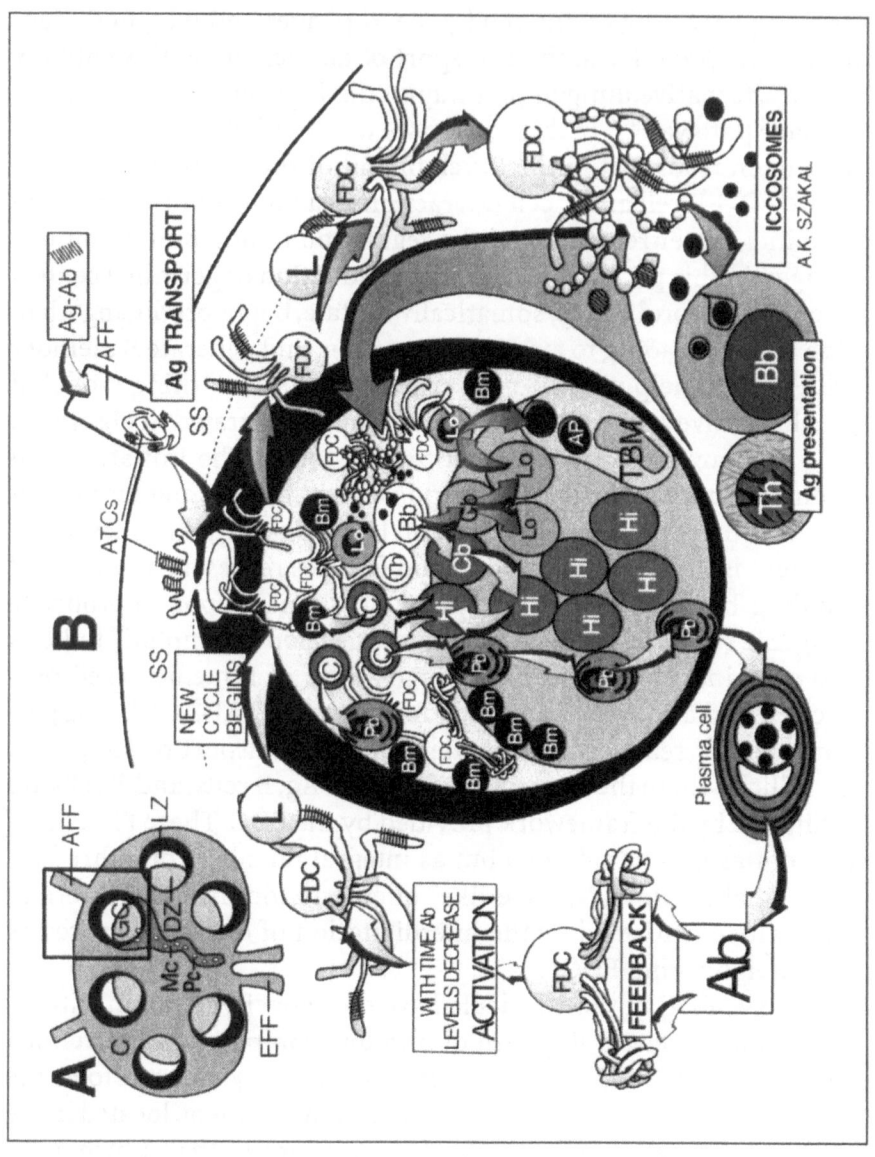

Fig. 2.1 (opposite). The alternative antigen transport pathway and sequence of immunological events. (A) (upper left corner) illustrates a midsagittal section of a lymph node depicting the location of secondary lymphoid follicles or nodules, germinal centers (GC), their light (LZ) and dark zones (DZ) and a plasma cell (Pc) laden medullary cord (Mc). The rectangle indicates the follicle shown enlarged in (B) with its cellular components and associated cellular events. The major events in the alternative antigen (Ag) transport, the transport of Ag-Ab complexes, the morphological changes and maturation of FDC occurring primarily in the light zone, are shown in sequence, enlarged, surrounding the lymphoid follicle. The Ag-Ab complexes binding to the dendrites of Ag transporting cells (ATC) and FDC were observed to form a spiraling pattern with a mean periodicity of 465 Å shown to favor binding of B cell Ag receptors. This is illustrated by the repeating parallel hatch marks on some of the dendrites.

The sequence of events in the pathway is the following: Ag transport—Ag-retaining FDC-reticulum development (1-24 hr); initial antigenic stimulation of specific B clones and/or contact with FDC for costimulation; iccosome formation and dispersion (day 1-3); iccosome endocytosis, Ag processing and presentation to Th cells by specific B cells (Bb) (day 3-5)—the large arrow indicates location of these events in the light zone. Next, there is a proliferative expansion of B cells in the dark zone. The B blasts come to be situated at the junction of the light zone and dark zone, begin dividing and give rise to the so called centroblasts (Cb). As the result of somatic hypermutation these Cb will be of varying affinities. There will be those with high affinity (Hi) and those with low affinity (Lo) for the Ag. Centroblasts are thought to migrate into the light zone where they may compete for antigenic binding sites on FDC. Those antigenically stimulated stay alive and become centrocytes (C), which may mature to form plasma blasts (Pb), or memory B cells (Bm). Plasma blasts subsequently migrate to the medullary cords (Mc) and bone marrow to become Ab producing plasma cells. Low affinity centroblasts will not compete successfully for antigenic binding sites and will die by apoptosis and are phagocytized by tingible body macrophages (TBM) located in the dark zone of lymph node germinal centers. As shown in the diagram, the Ab produced by plasma cells feeds back on the Ag still retained on FDC dendrites and crosslink them to hide the Ag-Ab complexes within the "ball-of-yarn"-like convolutions of the dendrites. This Ag-Ab equilibrium is maintained until Ab levels drop. At that time, Ab dissociates from some of the Ag on the dendrites, allowing them to unravel and again expose the Ag to Bm cells. This "activation process" then initiates a new cycle of events leading via Ab and Bm cell production to the maintenance of humoral immunity. Reprinted with permission from Tew JG et al, Immunol Rev 1997; 156:39-52. ©1997 Munksgaard International Publishers Ltd., Copenhagen, Denmark.

have C3 fragments associated and C3 fragments can reduce the dose of antigen required to elicit a response 1,000- to 10,000-fold.[19] This is thought to be a consequence of co-crosslinking sIg and CR2 (CD21) on the B cells.[19-21] Antigen injected subcutaneously into immune animals immediately complexes with specific Ab and complement would be fixed. The majority of Ag-Ab complexes are endocytosed by macrophages, rapidly degraded, and cleared from the system in 2-3 days.[22,23] However, a small fraction of the immune complexes are transported by a group of nonphagocytic cells with dendritic morphology to lymphoid nodules (follicles) in the draining lymph nodes.[22] These nonphagocytic cells transporting the antigen from the afferent lymph, through pores in the floor of the subcapsular sinus, to the follicles, are "monocyte-like" cells with immune complex coated veil-like processes. These cells have been described as penetrating "frilly" or "veiled" cells and we refer to these cells collectively as antigen transport cells (ATC).[22]

Antigen transport sites can be recognized within one minute after challenge of immune mice.[22] During transport, ATC form an immune complex coated chain between the subcapsular sinus and the prospective site of the FDC-reticulum in the follicles. Gradual morphological changes take place in ATC along this chain and they are morphologically similar to FDC near the site of the forming FDC-reticulum. This morphological transition[22] and shared surface determinants between ATC and FDC[24] prompted us to suggest that some ATC may be pre-FDC.[24,25]

Antigen transport is continuous with the developing antigen retaining reticulum and this provides the basis for the concept that one antigen transport site results in the development of one antigen retaining reticulum. Antigen transport cells coated with immune complexes appear at the ultrastructural level to bind follicular lymphocytes to their immune complex coat. This appears similar to FDC binding lymphocytes in the antigen retaining reticulum.[22] Studies with B cell depleted[26,27] and SCID[28] mice showed that B cells are required for FDC development. Injection of normal B cells, or to some extent T cells, into SCID mice facilitates some development of FDC-reticula, but full development of FDC-reticula are supported only by the reconstitution of both, B cell and T cell populations.[28]

As shown by ultrastructural studies,[29,30] FDC retain the complexed antigen on the surface of convoluted filament-like (filiform) dendritic processes. Some of these FDC mature and develop "beaded" dendrites which form immune complex coated bodies or iccosomes.[30] Iccosome formation requires the interaction of the beaded dendrites of FDC with pleomorphic dendrites (dendrites of irregular shapes) of another subpopulation of FDC. These FDC with pleomorphic dendrites help concentrate immune complexes on the surface of the beads. During this process some antigen also enters the interior of the iccosomes.[9,30] The iccosomes are dispersed around day three after antigenic challenge and are endocytosed by follicular B cells. The antigen is then processed in endocytic vesicles, becomes associated with Golgi lamellae and is reexpressed for presentation to T cells.[9,30,31] Thus, the combination of FDC-retained antigen,[32-34] iccosomes, FDC costimulation,[35,36] and T cell lymphokines, are involved in germinal center reactions.[9] The delivery of antigen in a concentrated form by the "palatable" (0.3-0.5 μm diameter) iccosome may represent an extension of the accessory activity of FDC.

Not all FDC take part in iccosome formation and antigen is retained within "ball of yarn"-like convolutions of dendritic processes[30] for long periods (months to years).[17] It is thought that this retained antigen plays an important role in the regulation and long term maintenance of humoral immunity.[23] This is further supported by the data presented below, where specific immune responses are not maintained in aged or retrovirus infected hosts, where the FDC fail to preserve the antigen.

The Effect of Aging on Antigen Transport, Iccosome Production, Germinal Center Formation, and Retention of Ag-Ab Complexes

Immune complexes are handled in old mice by the cells and mechanisms described above for young adult mice. However, a number of deficits have been noted in the way old mice handle these immune complexes and these are summarized in Table 2.1; some of these differences are illustrated in Figures 2.2 and 2.3. Antigen transport leads to the formation of the antigen retaining FDC-recticula (described above and illustrated in Fig. 2.3a). On average, in young adult mice 8 reticula develop per popliteal lymph node and 12 per axillary lymph node after challenge of passively immunized mice [anti-horse radish peroxidase (HRP) was used for passive immunization] with the histochemically detectable antigen HRP.[37,38] In marked contrast, in aged mice (given the same passive immunization) the number of reticula in the popliteal lymph node drop from 8 to 2 and from 12 to 4 in the axillary lymph node.[37,38] Furthermore, the reticula in the aged mice look smaller and are poorly developed when compared with reticula in young adult mice.[37,38] The combination of small numbers and small size results in a 10- to 100-fold decrease in the tissue volume occupied by the antigen bearing reticula in aged mice. This is illustrated in Figure 2.2 where popliteal and axillary lymph nodes in young adult mice and old mice are compared in terms of the kinetics of changes after antigen challenge, including changes in size of the lymph nodes, antigen retaining reticulum (ARR), and germinal centers. Note that the popliteal lymph nodes in old and young adult mice were about the same size at the time when the antigen challenge was given and that the old axillary lymph nodes were larger than those in the young adult mice. However, by 3 days

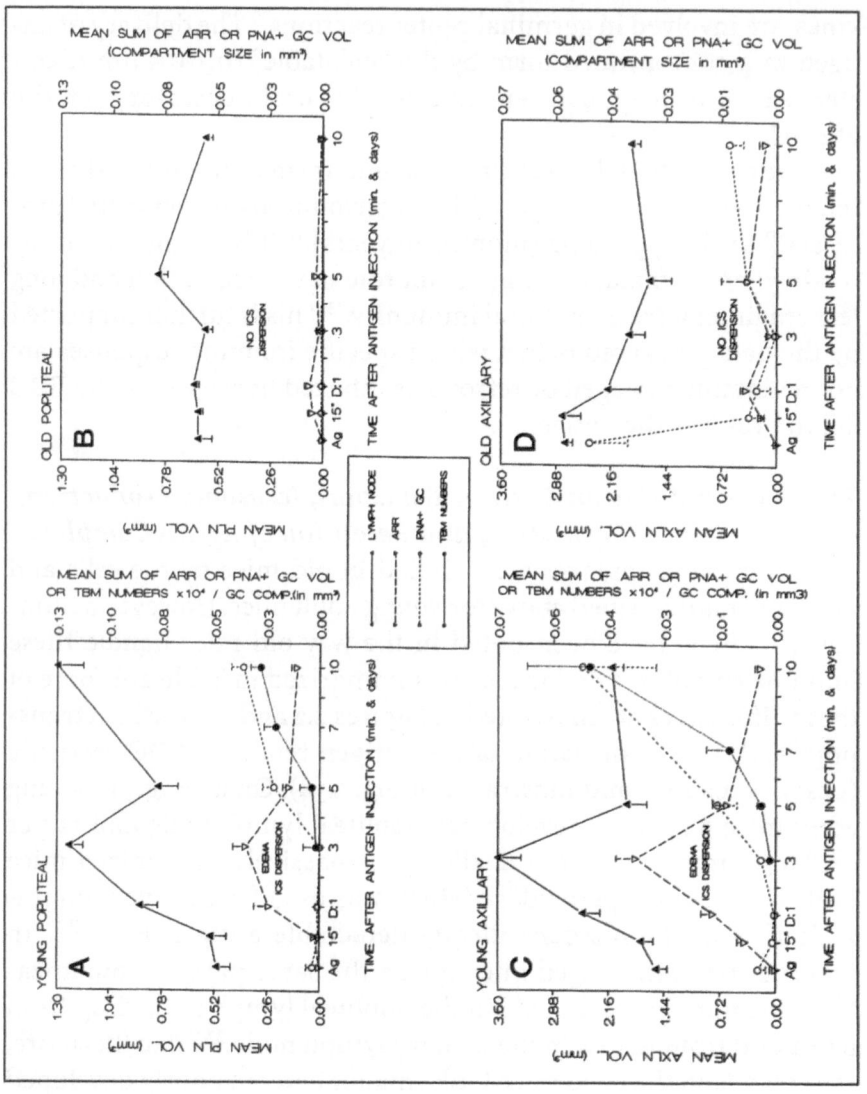

Fig. 2.2 (opposite). Charts illustrating comparisons of the kinetics of size changes in popliteal (PLN) and axillary (AXLN) lymph nodes, antigen retaining reticulum (ARR), and PNA positive germinal center (GC) compartments in young (6-8 weeks) and old (23 months) mice during the germinal center reaction. The kinetics show a direct relationship between ARR development (first phase) and dissipation and lymph node size changes induced by the challenge injection of the antigen HRP. A similar relationship exists between de novo germinal center development (second phase) and PLN size increase (A). In the AXLN, although the kinetics of lymph node size changes parallel that of the ARR, in the second phase (unlike in the PLN), the rate of lymph node size increase takes place at a much lower rate than the rate of increase in germinal center size (C). This is probably the result of the greater capacity of the larger AXLN. Note that iccosome dispersion, accompanied by local edema in the ARR, occurs at the time of peak development of the ARR in young adult mice (A, C). Iccosome dispersion was shown to result in the delivery of the antigen to B cells for processing and presentation to T cells at the initiation (day 3) of germinal center development.[30,31] The age-related defects in antigen transport, ARR development, and iccosome production are reflected in the depressed kinetic profile of germinal center development and lymph node size changes (B and D). Note the dissociation (size reduction and disappearance) of preexisting germinal centers between the time of antigen (Ag) challenge and day 3 is especially well illustrated in old AXLN which tend to have a large number of environmentally induced (preexisting) germinal centers (D) In old AXLN, preexisting germinal center dissociation is accompanied by a size reduction in AXLN size. Observe that this dissociation is nearly the reverse of de novo germinal center development in young adult AXLN. Error bars represent standard error of the mean. Reprinted with permission from Szakal AK et al, Anat Rec 1990; 227:475-485.

after challenge both the young adult popliteal and young adult axillary lymph nodes had responded to the challenge, were much larger and the space (compartment size) occupied by the ARR was 10 to 100 times larger than the ARR in the old mice (about 0.03 mm³ in young vs. < 0.003 mm³ in old). Note that it is around the day 3 period when the germinal centers become edematous and iccosomes (ICS in Fig. 2.2) are released. In short, the combination of reduced size and number of ARR results in a dramatic reduction of antigen on FDC in old mice and a reduction in the amount of immunogen available for release in the form of iccosomes for processing by the B cells.

The results in terms of germinal center development directly reflect the results with the ARR. Each ARR apparently leads to the development of one germinal center.[37,39] On average, in young adult mice the 8 reticula in the popliteal lymph node and 12 per axillary lymph node resulted in 8 and 12 germinal centers respectively.[38] In aged mice two reticula in the popliteal lymph node and 4 in the axillary lymph node resulted in 2 and 4 germinal centers.[38] Furthermore, the germinal centers in the aged mice look smaller and appear poorly developed when compared with those in

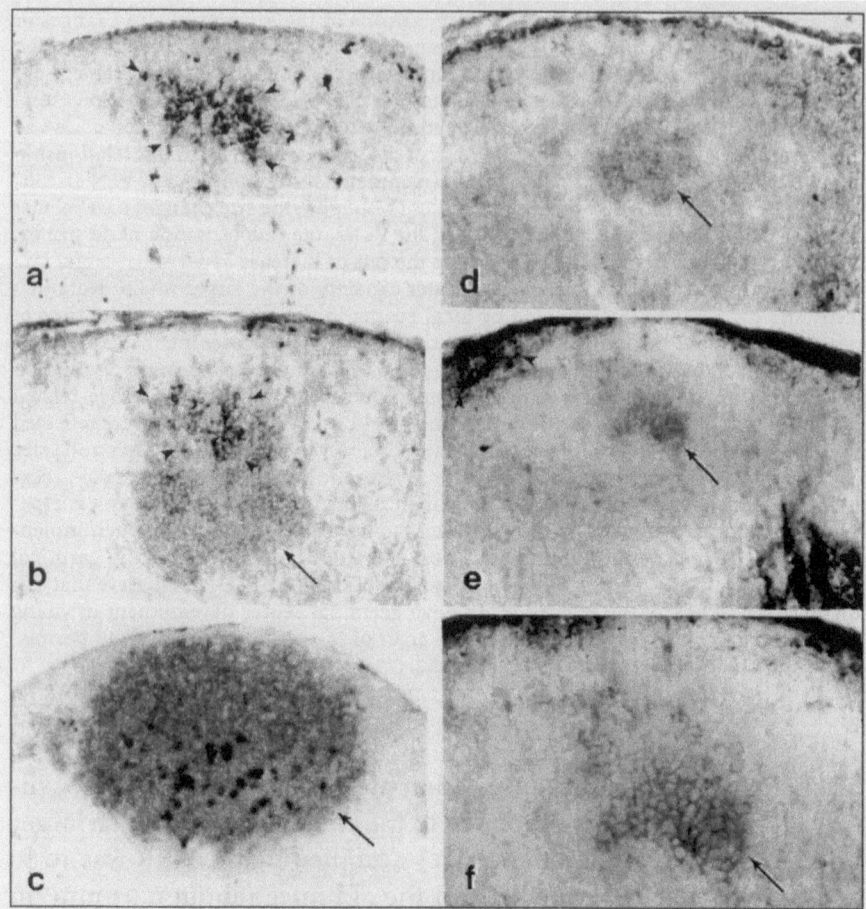

Fig. 2.3. Light micrographs illustrating the histochemical differentiation of preexisting and de novo PNA positive germinal centers in lymph nodes of mice passively immunized for the antigen, horseradish peroxidase (HRP).

(a) A peroxidase positive antigen retaining reticulum (ARR) of follicular dendritic cells (FDCs, arrowheads) at day 5 after footpad challenge with HRP as seen after histochemical localization of the trapped antigen.

(b) An adjacent section incubated with PNA-HRP conjugate for labeling germinal center B cells. Note the association of the peroxidase positive ARR (arrowheads), seen alone panel a, with the PNA positive germinal center (arrow) induced de novo by the HRP.

(c) A PNA positive germinal center induced the peroxidase negative antigen, human serum albumin (HSA) at day 5 after HSA challenge. Note the absence of a peroxidase positive ARR (MAC-2$^+$ tingible body macrophages intensely labeled with MAC-2 and glucose oxidase conjugate are also apparent in the lower half of the PNA positive germinal center).

(d) Popliteal, and (e) axillary preexisting PNA positive germinal centers (arrows) at 15 minutes after footpad challenge with HRP. Note that these germinal centers are small and are not associated with a peroxidase positive ARR (presumably induced by peroxidase negative environmental antigens). Also observe the HRP antigen transport site (arrowheads)[22,37] in panel (e), extending from the subcapsular sinus to the left of the lymphoid nodule.

the younger mice (see Fig. 2.3 for illustration). As with ARR, the combination of smaller numbers and size resulted in a decrease of 10- to 100-fold in germinal center volume in aged mice. Again this is graphically illustrated in Figure 2.2 where the germinal centers are largest at day 10 after antigen challenge in young adult mice. Note that the popliteal and the axillary lymph nodes in young adult mice were much larger and the space (compartment size) occupied by the germinal centers was about 0.04 mm^3 young vs. < 0.004 mm^3 in old. In contrast, in older mice the injection of antigen completely dissociated "preexisting germinal centers" (a previously described phenomenon[40]) in axillary lymph nodes within 3 days (see Fig. 2.2d) and only an occasional small de novo germinal center could be

Table 2.1. Abnormalities in FDC function in aged animals

Abnormality	Associated Features
1. Reduced number of FDC reticula	One FDC reticulum results in one GC Reduced reticula means reduced GC
2. FDC reticula are small	Small reticula give small GC and the combination of low number and small size gives a GC volume reduced by > 90%
3. Much of the Ag is not localized in the follicles	In contrast with the young animals, the Ag in old animals is associated with cells near the subcapsular sinus
4. FDC look atrophic	They appear smaller, with fewer processes, less Ag, and interact with fewer B cells
5. Iccosomes do not appear to be produced	Reduced Ag for GC B cells to process & AFC production is reduced
6. Level of retained Ag declines rapidly	In contrast with young mice that maintain Ag for many months or years, Ag in the aged animals disappears in a matter of a couple of weeks and serum Ab levels are not sustained

(f) The same PNA positive germinal center as that seen in e at a higher magnification to illustrate better the lack of a peroxidase positive ARR. Note that if this germinal center were induced by the HRP injection, then it should have had a peroxidase positive ARR associated with it staining about as intensely as the antigen transport site in (e). In this passively immunized mouse model, the association of a peroxidase positive ARR with a PNA positive germinal center was interpreted as proof of the de novo induction of the germinal center by HRP. Conversely, its lack indicated preexisting germinal centers. (a) and (b), x 165, c, x 125, (d) and (e), x 165; and (f), x 320 Reprinted with permission from Szakal AK et al, Anat Rec 1990; 227:475-485.

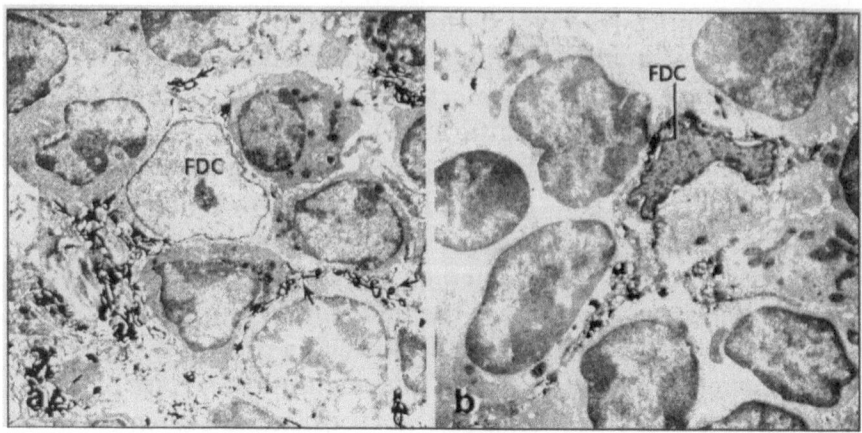

Fig. 2.4. Electron micrographs illustrating HRP localization on FDC in young adult and old mouse PLN. Panel a shows an FDC from a young PLN at day 5 with its large euchromatic nucleus and its highly convoluted, HRP-anti-HRP coated dendrites (arrows) among the large B lymphocytes of an ARR. Panel b illustrates an atrophic FDC from a defective ARR in an aged mouse. Note the shrunken nucleus containing an electron dense nucleoplasm and the electron dense dendritic processes associated with less HRP-anti-HRP than the dendrites of the young FDG in panel a. Panel a, X 3990 and b, X 5670. Reprinted with permission from Szakal AK et al, Aging: Immunol Inf Dis 1988; 1:7-22.

detected during the following 10 day period (Fig. 2.2d). In short, the combination of reduced ARR size and number correlated with a dramatic reduction in germinal center volume in old mice.

Studies at the ultrastructural level further support the belief that senescent FDC are functionally compromised.[37] The FDC were fewer in number and did not form the extensive convolutions typical of FDC in younger mice (see Fig. 2.4). The amount of electron dense material associated with their dendrites (immune complexes) was scanty.[37] The cell bodies appeared reduced in size and contained a condensed, more electron dense nucleoplasm. Instead of numerous dendrites emanating from the cell body, these FDC had one or two larger processes from which a few shorter dendritic processes extended among the lymphocytes.[37] It would appear that less of the antigen was exposed to the surrounding lymphocytes and this atrophic appearance become more exaggerated with time after antigen injection.

In addition, we were unable to find FDC in the aged mice producing iccosomes and we think this is a major reason why secondary antibody responses in old mice are so depressed. In an attempt to determine the role of defective (noniccosome producing) FDC in the age-related deficit, cell transfer experiments were done. The results showed that young adult controls with normal iccosome producing FDC gave normal anamnestic responses upon transfer of normal memory T and B cells. In contrast, in old mice, the anamnestic response could not be restored with young memory T and B cells.[4] Thus we reasoned that the depressed antibody response in aged mice is not simply attributable to old B and T cell problems. To help determine how important iccosome production was, iccosome-like fragments were produced by sonicating young and old FDC. When young adult memory T and B cells were preincubated with these iccosomal preparations and transferred into naive young adult irradiated recipients, secondary responses were obtained from either young or old FDC.[4] In contrast, if the FDC were not sonicated to force iccosome production, only the young iccosome producing FDC gave good antibody responses.[4] These results support the concept that FDC are required for the anamnestic response and that the defective FDC are not replaceable simply by providing young T and B cells. The data do not exclude the possibility that defects are present in aged lymphocytes, but the B cell subpopulation identified as being $J11D_{Lo}$ by use of an appropriate monoclonal antibody is not diminished in number in aged mice.[41] This is important because these cells give rise to secondary antibody responses[42] and are the B cells involved in germinal center development.[43] Further, the results described above suggest that iccosomes or iccosome-like fragments are necessary vehicles for the delivery of the FDC retained antigen to B cells for the development of a successful secondary antibody response.

In young adult animals antigen is retained for months to years on FDC and the secondary response is likewise maintained and typically increases in affinity as the response matures.[23] In contrast, in the old mice much of the antigen retained for 5 to 10 days is not on the FDC[37,44] and the vast majority of the antigen retained by the FDC has disappeared in about 10 days.[37] Interestingly, the specific antibody responses are not sustained in old mice[4] and this is consistent with the loss of retained antigen.

The Effect of Retrovirus Infection on Antigen Transport, Iccosome Production, Germinal Center Formation, and Retention of Ag-Ab Complexes

In 1968, the colocalization of exogenous antigen with a murine retrovirus in the FDC reticulum was observed by Szakal and Hanna.[45] A subsequent report by these authors provided the first ultrastructural demonstration of the trapping of exogenous retrovirus particles (Rausher murine leukemia virus) by FDC.[46] The beginning of the AIDS epidemic, and the initial reports by Armstrong and Horne[47] and Tenner-Racz et al[48] on HIV trapping by FDC, focused attention on the involvement of FDC in the development of persistent generalized lymphadenopathy and AIDS. Histopathological analysis by a number of workers indicates progressive changes in the lymphoid follicles of infected mice and men including: (1) marked nodal enlargement with florid follicular hyperplasia;[47-51] (2) involution characterized by mixed follicular hyperplasia and follicular fragmentation;[47-52] and (3) nodal regression with severe follicular atrophy and loss of FDC.[47-50]

Table 2.2 identifies a number of abnormal/atypical characteristics of FDC and germinal centers in retrovirus infected animals. Most antigens are not found on FDC until an antibody response has been induced and immune complexes form. However, some antigens are known to be able to activate the complement cascade and localize on FDC before antibody is formed.[1] This is likely the case with retroviruses. HIV is reported to activate the complement system through gp 41,[53] and retroviruses are found associated with FDC within a short period after infection (we have seen them within 24 hours).[54-56] Almost immediately after the germinal centers begin to form, they become both numerous and hyperplastic,[47,48] and this corresponds with the lymphadenopathy characteristic of retrovirus infections. The virus appears to alter normal FDC function and dominate the cell. Antigens that were trapped by the FDC prior to infection come off, suggesting that the virus is displacing these immune complexes.[56] Furthermore, when new antigen is injected it is not trapped by the FDC in spite of the fact that the FDC are present in large numbers.[56] In short, the virus appears to monopolize the ARR development.

A remarkable feature of the FDC-HIV association is that the virus remains intact and infectious.[8,57] As HIV on FDC is in the form of immune complexes, we reasoned that it would be difficult for the virus to infect T cells. To our surprise, HIV on FDC is highly infectious. Furthermore, FDC can convert neutralized HIV into an infectious form even in the presence of a vast excess of neutralizing antibody.[8,57] Thus FDC may provide a mechanism whereby HIV infection can continue in the presence of neutralizing antibody which may be present during the long period of clinical latency. Interestingly, it is known that active viral infection occurs primarily in lymphoid fol-

Table 2.2. Abnormalities in FDC Function in Retrovirus infected Animals

Abnormality	Associated Features
1. Retrovirus is rapidly trapped	Hyperplastic GC form rapidly and are numerous
2. Virus begins to monopolize the retention system	Soluble Ag trapped prior to infection comes off and new Ag can not be loaded on FDC
3. Virus are retained intact	The retained virus maintains its ability to infect activated T cells and can no longer be blocked by specific neutralizing Ab
4. Virus are retained for long periods	Soluble Ag are retained for months to years Similarly, the retrovirus are not only retained but maintain infectivity for months in GC micro-environment
5. FDC are destroyed in later stages	Follicles lyse, architecture of the secondary lymphoid tissue disappears. Some PNA positive sites are present but no FDC, light zones, dark zones, etc.

licles during the long period of clinical latency[58,59] and we reason that the large reservoir on the FDC is likely responsible for this continuing infection.[8,57]

It is known that antigens are retained intact in the immune complexes on FDC for months to years. It seemed reasonable that retroviruses like HIV would also be retained. However, HIV is highly labile and has a half-life of only a few hours both in vivo[60-62] and in vitro (unreported observation). Consequently, we reasoned that the virus might be retained by FDC but would rapidly lose infectivity. Remarkably, we have found, in experiments that are continuing as this report is being written, that HIV loaded onto murine FDC remains infectious for months.[57,63] The mouse is a nonpermissive host and the virus is not replicating in this animal; however, when the HIV-bearing-FDC from the mouse are isolated and co-cultured with human T cells, the T cells become infected.[8,57] This phenomenon can be demonstrated for months after loading virus on FDC by injecting the mouse with anti-HIV followed by HIV. We reason that treatment of patients with anti-retrovirus agents is not likely to inactivate virus which is not replicating on the surface of the FDC. Thus, we reason that new infection may take place repeatedly in human lymphoid follicles and may replenish the FDC reservoir whenever the inhibitory therapy is interrupted. It may be important to inactivate the virus on the FDC in patients being treated with anti-viral drugs to minimize the potential for continual reinfection by virus in this reservoir. The mechanism by which FDC maintain virus infectivity are not understood but are under study at present.

Ultimately, FDC are destroyed in both mouse, man, and non-human primates.[49,50,64,65] This typically occurs toward the end of clinical latency and the immune system at that point is dysfunctional. Destruction of the FDC is associated with follicular lysis and a loss of normal architecture in the secondary lymphoid tissues.[49,50,64] Interestingly, some foci of PNA positive areas persist but these are not germinal centers with light zones, dark zones, FDC, tingible body macrophages, etc.[50] The mechanism for destruction of the FDC is not clear. It was suggested very early after the discovery of the virus in association with FDC that the FDC may be infected and destroyed, and a number of published reports support the view that FDC are infected.[49,50,64] However, this issue remains controversial. It has also been reported in both man and macaques that there is an influx of CD8[+] lymphocytes into germinal centers and interfollicular areas and that these cells may contribute to the destruction of FDC via their cytotoxic activity.[65-67] These cells possess granzyme B and are present at a 20-fold increase over that observed in noninfected individuals. We reason that virus particles trapped on the surface of FDC may serve as a target for CD8[+] lymphocytes by virtue of MHC class I molecules (on the viral envelope) bearing HIV peptides obtained from host cells during the budding process. Although the mechanism of FDC destruction has not been elucidated, it is clear that the FDC network is eventually destroyed as a consequence of the retrovirus.

FDC Mediated Stimulation of B Cell Functions, Initiation of Specific Ab Production in Germinal Centers and Ab Production in Bone Marrow

The germinal center is recognized as a center for production of memory B cells. However, cells of the plasmacytic series are also produced.[14,15,68,69] The number of Ab forming cells (AFC) in germinal centers peaks during an early phase of the germinal center reaction (3 to 5 days after secondary antigen challenge) and then declines. During this early phase, germinal center B cells receive signals needed to become AFC. The germinal center becomes edematous and the AFC leave and we find them in the thoracic duct lymph and in the

Table 2.3. Characteristics of antibody response in aged mice and men

Feature	Associated Phenomenon
1. Serum immunoglobulin levels are high even elevated	Clearly B cells are capable of being stimulated, differentiating, and making IgG.
2. Specific Ab responses are markedly depressed	It may be difficult to get good protective specific immunity to infectious agents
3. Ag bearing FDC do not elicit specific Ab responses, however, sonicating FDC to release iccosomes results in Ab responses	Young FDC and sonicated old FDC will give Ab. The data suggest that old FDC have a basic problem in producing iccosomes.

blood.[15] These germinal center AFC home to the bone marrow where they mature and produce the vast majority of Ab in secondary responses.[15,69] In the second phase, which peaks about 10 days after challenge, germinal centers enlarge and the memory B cell pool is restored and expanded and there are very few AFC present.[14]

The Effect of Aging on FDC Mediated Stimulation of B Cell Functions

In aged mice and men the level of serum immunoglobulins are normal or even elevated,[70-72] suggesting that the ability of the B cells to make immunoglobulin molecules is reasonably normal (Table 2.3). However, attempts to immunize aged animals with simple protein antigens typically result in poor responses.[4,5] Furthermore, memory is difficult to detect and secondary responses are not much improved over primary responses.[4] Secondary antibody responses are characterized by long-lasting production of antigen specific IgG that may provide protection against infectious diseases for long periods of time.[6,7] Thus, the inability to get a good secondary response and sustain these responses may place aged individuals at increased risk for a variety of infectious diseases.

We reasoned that a major problem in eliciting and maintaining recall responses in the aged may be with the FDC dependent antigen handling mechanisms. To begin testing, we transferred normal memory T and B cells into aged mice and normal mice (positive control) and then challenged with specific antigen. Young adult mice given the T and B memory cells gave a secondary response upon antigen challenge, but antigen in aged animals, with the defective

FDC, failed to stimulate the young memory cells.[4] Furthermore, young antigen-bearing FDC cultured with young memory T and B cells elicited a specific antibody response. In contrast, incubating aged antigen-bearing FDC with young memory T and B cells failed to elicit a comparable response.[4] Since iccosome formation has not been observed in aged mice, iccosome-like fragments were generated by sonicating FDC from aged mice and these were tested for their ability to induce an anamnestic response. This procedure restored the ability of antigen retained on the FDC from aged mice to induce a normal anamnestic response.[4] These data support the concept that the inability to form and disperse iccosomes contributes to the impaired ability of aged mice to mount anamnestic antibody responses and provides further support for the role of FDC and iccosome in anamnestic responses as part of the early phase of the germinal center reaction.[4]

The Effect of Retrovirus Infection on FDC Mediated Stimulation of B Cell Functions

In retrovirus infected mice and men, the level of serum immunoglobulins are normal or even elevated through the period of clinical latency and beyond (Table 2.4).[73,74] This suggests that the ability of the B cells to make immunoglobulin molecules is reasonably normal through this period. However, attempts to immunize retrovirus infected people and animals with simple protein antigens often result in poor primary responses and secondary responses that are not much improved over the primary responses.[4,56] This inability to induce and maintain a high titer of high affinity antibody may, in part, relate to the increased risk for a variety of infectious diseases in HIV infected patients.

As explained above, FDC and associated retained antigen are thought to be involved in the maintenance of serum levels of specific IgG with high affinity. Accordingly, we reasoned that the loss of retained antigen on the FDC should correlate with reduction in the level of specific antibody maintained in the serum. To test this, sera from retrovirus infected and control mice were used to study the levels of specific IgG persisting at various times after infection. Antigen trapped on the FDC prior to infection began to disappear around the second and third week of infection and was practically gone by six weeks.[50] Specific antibody titers in the serum were maintained normally for about three weeks after infection but then declined precipitously. By six weeks after infection the antibody levels

Table 2.4. Characteristics of antibody response in retrovirus infected mice and men

Feature	Associated Phenomenon
1. Serum Ab is high, even elevated	Clearly B cells are capable of being stimulated, differentiating, and making IgG
2. Specific Ab responses are markedly depressed	It is difficult to get good protective specific immunity to infectious agents
3. Specific Ab to a variety of soluble Ag declines	The patients become more susceptible to opportunistic infections
4. Antibody specific for p24 declines over time	The decline in anti-p24 may relate to inability to maintain soluble Ag on FDC
5. Some specific antibodies may persist (e.g., anti-env)	The intact virus appears to monopolize FDC and the epitopes on viral-env may be available to maintain a stimulus

were markedly depressed relative to normal.[50] FDC were present in this period two to three weeks after infection and they appeared normal; however, specific antigen was apparently being displaced by the virus and new antigen could not be added to the FDC reticulum.[50] The obvious presence of FDC during this early period indicate that retroviral infections may cause FDC dysfunctions long before FDC are destroyed or even appear damaged. In short, the FDC appear to be monopolized by the virus and contribute to the lymphadenopathy so prominent during this period by eliciting the hyperplastic germinal centers. The problem is that the these FDC fail to handle normal antigens and support normal immune function.

Concluding Comments

Most attempts to understand the lack of immunological responsiveness in the aged and retrovirus infected individuals have focused on lymphocytes. We reasoned that a major problem in eliciting normal germinal centers, inducing immunological memory, and maintaining recall responses with high affinity antibody may be attributable, at least in part, to a problem with FDC. As discussed here, FDC in aged animals were defective and iccosome formation was not observed. However, when iccosome-like fragments were artificially generated by sonicating FDC from aged mice they were able to induce a normal anamnestic response. These data support the concept that the inability to form and disperse iccosomes contributes to the impaired ability of aged mice to mount anamnestic antibody responses and provides further support for the role of FDC and iccosomes in anamnestic responses as part of the early phase of the germinal center reaction. After retrovirus infection the FDC appear to be monopolized by the virus and contribute to the lymphadenopathy by eliciting hyperplastic germinal centers. The problem is that the these FDC fail to handle normal antigens and support normal immune function. Specific antigens trapped prior to infection were apparently being displaced by the virus and new antigen could not be added to the FDC reticulum. Consequently, it is not surprising that it is difficult to induce and maintain specific antibody responses with high affinity. In short, FDC and associated function is abnormal in aged and retrovirus infected individuals. We reason that reestablishment of FDC function may help restore immunological memory and secondary antibody responses with high affinity antibody. Restoring these recall responses could be critical to establishing and maintaining immunity to a variety of infectious agents in the aged and retrovirus infected. However, we are also aware that the FDC trap retroviruses and that the viruses retained on the FDC may be highly infectious.[8] This may complicate efforts to restore immune function in retrovirus infected hosts, but this also suggests that it is critical to understand FDC and FDC functions in the regulation of humoral immune responses.

Acknowledgments

This research was supported by Grants AG-05379, AI-17142, AI-34631, AI-32406 and AI-39963 from the National Institutes of Health.

References

1. Bernstein LJ, Ochs HD, Wedgwood RJ et al. Defective humoral immunity in pediatric acquired immune deficiency syndrome. J Pediatr 1985; 107:352-357.
2. Aucouturier P, Couderc LJ, Gouet D et al. Serum immunoglobulin G subclass dysbalances in the lymphadenopathy syndrome and acquired immune deficiency syndrome. Clin Exp Immunol 1986; 63:234-240.
3. Sirianni MC, Rossi P, Scarpati B et al. Immunological and virological investigation in patients with lymphoadenopathy syndrome and in a population at risk for acquired immunodeficiency syndrome (AIDS), with particular focus on the detection of antibodies to human T-lymphotropic retroviruses (HTLV III). J Clin Immunol 1985; 5:261-268.

4. Burton GF, Kosco MH, Szakal AK et al. Iccosomes and the secondary antibody response. Immunology 1991; 73:271-276.
5. Szakal AK, Kapasi ZF, Masuda A et al. Follicular dendritic cells in the alternative antigen transport pathway: Microenvironment, cellular events, age and retrovirus related alterations. Semin Immunol 1992; 4:257-265.
6. Gray D, Leanderson T. Expansion, selection, and maintenance of memory B cell clones. Curr Top Micrbiol Immunol 1990; 159:2
7. Cerottini JC, MacDonald HR. The cellular basis of T-cell memory. Annu Rev Immunol 1989; 7:77-89.
8. Heath SL, Tew JG, Szakal AK et al. Follicular dendritic cells and human immunodeficiency virus infectivity. Nature 1995; 377:740-744.
9. Szakal AK, Kosco MH, Tew JG. Microanatomy of lymphoid tissue during the induction and maintenance of humoral immune responses: structure function relationships. Annu Rev Immunol 1989; 7:91-109.
10. Tew JG, Kosco MH, Szakal AK. The alternative antigen pathway. Immunol Today 1989; 10:229-231.
11. Szakal AK, Burton GF, Smith JP, Tew JG. Antigen processing and presentation in vivo. In: Spriggs DR, Koff WC, eds. Topics in Vaccine Adjuvant Research. Boca Raton: CRC Press, 1991:11-23.
12. Tew JG, Kosco MH, Burton GF et al. Follicular dendritic cells as accessory cells. Immunol Rev 1990; 117:185-211.
13. Tsiagbe VK, Inghirami G, Thorbecke GJ. The physiology of germinal centers. Crit Rev Immunol 1996; 16:381-421.
14. Kosco MH, Burton GF, Kapasi ZF et al. Antibody-forming cell induction during an early phase of germinal centre development and its delay with ageing. Immunology 1989; 68:312-318.
15. DiLosa RM, Maeda K, Masuda A et al. Germinal center B cells and antibody production in the bone marrow. J Immunol 1991; 1460: 4071-4077.
16. MacLennan IC. Germinal centers. Annu Rev Immunol 1994; 12:117-139.
17. Tew JG, Mandel TE. Prolonged antigen half-life in the lymphoid follicles of specifically immunized mice. Immunology 1979; 37:69-76.
18. Mitchell J, Abbot A. Ultrastructure of the antigen-retaining reticulum of lymph node follicles as shown by high resolution autoradiography. Nature 1965; 208:500-502.
19. Dempsey PW, Allison ME, Akkaraju S et al. C3d of complement as a molecular adjuvant: Bridging innate and acquired immunity. Science 1996; 271:348-350.
20. Small TN, Keever CA, Weiner-Fedus S et al. B-cell differentiation following autologous, conventional, or T-cell depleted bone marrow transplantation: A recapitulation of normal B-cell ontogeny. Blood 1990; 76:1647-1656.
21. Tew JG, Wu J, Qin D et al. Follicular dendritic cells and presentation of antigen and costimulatory signals to B cells. Immunol Rev 1997; 156:39-52.

22. Szakal AK, Holmes KL, Tew JG. Transport of immune complexes from the subcapsular sinus to lymph node follicles on the surface of nonphagocytic cells, including cells with dendritic morphology. J Immunol 1983; 131:1714-1727.

23. Tew JG, Phipps RP, Mandel TE. The maintenance and regulation of the humoral immune response: Persisting antigen and the role of follicular antigen-binding dendritic cells as accesory cells. Immunol Rev 1980; 53:175-201.

24. Haley ST, Tew JG, Szakal AK. The monoclonal antibody FDC-M1 recognizes possible follicular dendritic cell precursors in the blood and bone marrow. Adv Exp Med Biol 1995; 378:289-291.

25. Tew JG, Mandel TE, Phipps RP et al. Tissue localization and retention of antigen in relation to the immune response. Am J Anat 1984; 170:407-420.

26. MacLennan ICM, Gray D. Antigen-driven selection of virgin and memory B cells. Immunol Rev 1986; 91:61-85.

27. Cerny A, Zinkernagel RM, Groscurth P. Development of follicular dendritic cells in lymph nodes of B-cell-depleted mice. Cell Tissue Res 1988; 254:449-454.

28. Kapasi ZF, Burton GF, Shultz LD et al. Induction of functional follicular dendritic cell development in severe combined immunodeficiency mice. Influence of B and T cells. J Immunol 1993; 150:2648-2658.

29. Szakal AK, Gieringer RL, Kosco MH et al. Isolated follicular dendritic cells: Cytochemical antigen localization, Nomarski, SEM, and TEM morphology. J Immunol 1985; 134:1349-1359.

30. Szakal AK, Kosco MH, Tew JG. A novel in vivo follicular dendritic cell-dependent iccosome-mediated mechanism for delivery of antigen to antigen-processing cells. J Immunol 1988; 140:341-353.

31. Kosco MH, Szakal AK, Tew JG. In vivo obtained antigen presented by germinal center B cells to T cells in vitro. J Immunol 1988; 140:354-360.

32. Mandel TE, Phipps RP, Abbot A et al. The follicular dendritic cell: Long term antigen retention during immunity. Immunol Rev 1980; 53:29-59.

33. White RG, French VI, Stark JM. A study of the localization of a protein antigen in the chicken spleen and its relation to the formation of germinal centers. J Med Microbiol 1970; 3:65-83.

34. Nossal GJV, Abbot A, Mitchell J et al. Antigen in immunity. XV. Ultrastructural features of antigen capture in primary and secondary lymphoid follicles. J Exp Med 1968; 127:277-290.

35. Burton GF, Conrad DH, Szakal AK et al. Follicular dendritic cells (FDC) and B cell co-stimulation. J Immunol 1993; 150:31-38.

36. Wu J, Qin D, Burton GF et al. Follicular dendritic cell (FDC) derived Ag and accessory activity in initiation of memory IgG responses in vitro. J Immunol 1996; 157:3404-3411.

37. Szakal AK, Taylor JK, Smith JP et al. Morphometry and kinetics of antigen transport and developing antigen retaining reticulum of fol-

licular dendritic cells in lymph nodes of aging immune mice. Aging: Immunol Inf Dis 1988; 1:7-22.

38. Szakal AK, Taylor JK, Smith JP et al. Kinetics of germinal center development in lymph nodes of young and aging immune mice. Anat Rec 1990; 227:475-485.

39. Szakal AK, Kosco MH, Smith JP, Tew JG. Kinetic and ultrastructural aspects of the antigen transport-FDC-Iccosome-B cell axis. In: Racz P, Tenner-Racz K, Dijkstra CD, eds. Morphological and Functional Aspects of Accessory Cells in Retroviral Infections. Basel, Switzerland: Karger-Verlag, 1991:29-43.

40. Hanna MG Jr, Nettesheim P, Ogden L et al. Reduced immune potential of aged mice: Significance of morphologic changes in lymphatic tissue. Pro Soc Exp Biol Med 1967; 125:882-886.

41. Linton PJ. The status of progenitors of memory B cells in aged mice. Aging: Immunol Inf Dis 1993; 4:35

42. Linton PJ, Decker DJ, Klinman NR. Primary antibody-forming cells and secondary B cells are generated from separate precursor subpopulations. Cell 1989; 59:1049

43. Linton PJ, Lo D, Lai L et al. Among naive precursor cell subpopulations only progenitors of memory B cells originate germinal centers. Eur J Immunol 1992; 22:1293-1297.

44. Holmes KL, Schnizlein CT, Perkins EH, Tew JG. The effect of age on antigen retention in lymphoid follicles and in collagenous tissue of mice. Mech-Ageing-Dev 1984; 25:243-255.

45. Szakal AK, Hanna MG Jr. The ultrastructure of antigen localization and virus-like particles in mouse spleen germinal centers. Exp Mol Pathol 1968; 8:75-89.

46. Hanna MG Jr, Szakal AK, Tyndall RL. Histoproliferative effect of Rauscher leukemia virus on lymphatic tissue: Histological and ultrastructural studies of germinal centers and their relationship to leukemogenesis. Cancer Res 1970; 30:1748-1763.

47. Armstrong JA, Horne R. Follicular dendritic cells and virus-like particles in AIDS-related lymphadenopathy. Lancet 1984; 2:370

48. Tenner-Racz K, Racz P, Dietrich M et al. Altered follicular dendritic cells and virus-like particles in AIDS and AIDS-related lymphadenopathy. Lancet 1985; 1:105-106.

49. Wood GS. The immunohistology of lymph nodes in HIV infection: A review. Prog AIDS Pathol 1990; 2:25-32.

50. Masuda A, Burton GF, Szakal AK et al. Loss of follicular dendritic cells in murine-acquired immunodeficiency syndrome. Lab Invest 1995; 73:1-10.

51. Biberfeld P, Porwit A, Biberfield G et al. Lymphadenopathy in HIV (HTLV-III LAV) infected subjects: The role of virus and follicular dendritic cells. Cancer Detect Prev 1988; 12:217-224.

52. LeTourneau A, Audouin J, Aubert JP et al. Viral type particles in the germinal centers during a lymphadenopathic syndrome related to AIDS. Ann Pathol 1985; 5:137-142.

53. Ebenblicher CF, Thielens NM, Vornhagen R et al. Human immuno-deficiency virus type 1 activates the classical pathway of complement by direct C1 binding through specific sites in the transmembrane glycoprotein gp41. J Exp Med 1991; 174:1417

54. Hurtrel B, Chakrabarti L, Hurtrel M et al. Early events in lymph nodes during infection with SIV and FIV. Res Virol 1994; 145:221-227.

55. Bach JM, Hurtrel M, Chakrabarti L et al. Early stages of feline immunodeficiency virus infection in lymph nodes and spleen. AIDS Res Hum Retroviruses 1994; 10:1731-1738.

56. Masuda A, Burton GF, Fuchs BA et al. Follicular dendritic cell function and murine AIDS. Immunology 1994; 81:41-46.

57. Burton GF, Masuda A, Heath SL, Smith BA, Tew JG, Szakal AK. Follicular dendritic cells (FDC) in retroviral infection: Host/pathogen perspectives. Immunological Rev 1997; 156:185-197.

58. Pantaleo G, Graziosi C, Demarest JF et al. HIV infection is active and progressive in lymphoid tissue during the clinically latent stage of disease. Nature 1993; 362:355-358.

59. Embretson J, Zupancic M, Ribas JL et al. Massive covert infection of helper T lymphocytes and macrophages by HIV during the incubation period of AIDS. Nature 1993; 362:359-362.

60. Perelson AS, Neumann AU, Markowitz M et al. HIV-1 dynamics in vivo: Virion clearance rate, infected cell life-span, and viral generation time. Science 1996; 271:1582-1586.

61. Wei X, Ghosh SK, Taylor ME et al. Viral dynamics in human immunodeficiency virus type 1 infection. Nature 1995; 373:117-122.

62. Ohmori H, Yamamoto I. Mechanism of augmentation of the antibody response in vitro by 2-mercaptoethanol in murine lymphocytes. Cell Immunol 1983; 79:173-185.

63. Smith BA, Tew JG, Szakal AK, Gartner S, Burton GF. Follicular dendritic cell maintenance of HIV infectivity. AIDS Pathogenesis 1997; 252(Abstr.).

64. Frost SD, McLean AR. Germinal centre destruction as a major pathway of HIV pathogenesis. J Acquir Immune Defic Syndr 1994; 7:236-244.

65. Rosenberg YJ, Zack PM, Leon EC et al. Immunological and virological changes associated with decline in CD4/CD8 ratios in lymphoid organs of SIV-infected macaques. AIDS Res Hum Retroviruses 1994; 10:863-872.

66. Tenner-Racz K, Racz P, Thome C et al. Cytotoxic effector cell granules recognized by the monoclonal antibody TIA-1 are present in $CD8^+$ lymphocytes in lymph nodes of human immunodeficiency virus-1-infected patients. Am J Pathol 1993; 142:1750-1758.

67. Koopman G, Wever PC, Ramkema MD et al. Expression of granzyme B by cytotoxic T lymphocytes in the lymph nodes of HIV-infected patients. AIDS Res Hum Retroviruses 1997; 13:227-233.

68. Sordat B, Sordat M, Hess MW et al. Specific antibody within germinal center cells of mice after primary immunization with horseradish peroxidase: A light and electron microscopic study. J Exp Med 1970; 131:77-91.

69. Tew JG, DiLosa RM, Burton GF et al. Germinal centers and antibody production in bone marrow. Immunol Rev 1992; 126:1-14.
70. Makinodan T, Kay MMB. Age influence on the immune system. Adv Immunol 1980; 29:287-330.
71. Doggett DL, Chang M, Makinodan T et al. Cellular and molecular aspects of immune system aging. Mol Cell Biochem 1981; 37:137-156.
72. Legge JS, Austin CM. Antigen localization and the immune response as a function of age. Aust J Exp Biol Med Sci 1968; 46:361-365.
73. Schnittman SM, Lane HC, Higgins SE et al. Direct polyclonal activation of human B lymphocytes by the acquired immune deficiency syndrome virus. Science 1986; 233:1084-1086.
74. Yarchoan R, Broder S. Immunology of HIV infection. In: Paul WE, ed. Fundamental Immunology. New York: Raven Press, 1989: 1059-1079.

CHAPTER 3

T and B Lymphocytes in Germinal Centers

R.A. Insel and M.H. Nahm

Germinal Center B Cells

General Overview of Germinal Center Reactions

As detailed in chapter 1, the germinal center can be histologically divided into a dark zone of proliferating centroblasts and a light zone of nonproliferating centrocytes.[1-3] The dark zone, in proximity to the PALS, is populated by proliferating centroblasts that clonally expand with a cell cycle time as short as 6-8 hours.[4] Up to 10 B cell clones populate a germinal center, but through selection there is an outgrowth of only 3-5 of these clones. The signals that activate migration from the PALS of an activated B cell to populate a primary follicle and form a germinal center reaction are incompletely elucidated and discussed further in chapter 1. Also, the mechanism(s) of the rapid clonal expansion of centroblasts in germinal centers is unknown, but discussed further below. These rapidly dividing centroblasts undergo a process of somatic hypermutation of the V regions of their immunoglobulin genes, detailed in chapter 4. Centroblasts develop into nonproliferating centrocytes that move apically into the basal light zone and then to the apical light zone where they are interspersed with a network of follicular dendritic cells (FDC) and $CD4^+$ T cells. It is in the light zone that centrocytes undergo massive apoptosis, with collection of condensed chromatin fragments as tingible bodies in macrophages, or escape apoptosis by a process of antigenic selection to develop into memory B cells or differentiate to become plasma cells. Germinal center B cells are programmed to die by apoptosis unless they are rescued by a positive signal. One

The Biology of Germinal Centers in Lymphoid Tissue, edited by G. Jeanette Thorbecke and Vincent K. Tsiagbe. © 1998 Springer-Verlag and R.G. Landes Company.

such signal comes from the interaction between germinal center B cells and FDC. The primary signal between the two cells in vivo is believed to be dependent on interaction with antigen held as immune complexes on the surface of the FDC. This interaction may facilitate a transfer of native Ag from the surface of the FDC as an immune complex-coated body (iccosomes) to the B cell, where it may be internalized, processed, and presented to T cells. The activated B cell can present processed Ag on its surface to the T cell and activate cytokine production from T cells. In vitro interactions of FDC with B cells leads to upregulation of MHC class II and, in the presence of Ag, upregulation of B7-2. (CD86) on the B cell. The germinal center response tapers by 3 weeks after immunization.

Precursors of Germinal Center B Cells

Both IgD$^+$ and IgD$^-$ B lymphocytes have been shown to give rise to germinal center formation.[5-7] Although sIgD expression on B lymphocytes is not obligatory, germinal center formation and memory B cell production may be enhanced by IgD$^+$ B cells through interaction with T cells that express an IgD receptor.[8] Furthermore, the presence of sIgD may interfere with the induction of tolerance and promote affinity maturation of the B cell.[9,10] Based on reconstitution of SCID mice, J11D^{lo} and J11D^{hi} splenic B cells differ in their potential to form germinal centers, with the former proving to be more efficient than the latter.[11] CD5$^+$ B cells are also poor at reconstitution of germinal centers.[11,12]

It has been suggested that human germinal center founder B cells express surface IgM and IgD and the germinal center markers CD38, CD10, CD71, and CD95/Fas, and have lost expression of Bcl-2 protein, which may account for their tendency to undergo spontaneous apoptosis in vitro.[13] Cells with this phenotype were found in the dark zone of the germinal center and lacked the CD23 and CD44 markers expressed on follicular mantle naive B cells. At least half of the cells in this subset contain germline encoded V genes, suggesting somatic mutation had not begun in some of these cells. However, mutation was observed in the other half of this subpopulation. These observations and the finding of somatically mutated V gene rearrangements in the absence of germinal centers in both lymphotoxin gene targeted knockout mice[14] and in CD40 ligand deficient children with the hyper-IgM syndrome[15] raise the question when after B cell activation, and where (PALS vs. germinal center) the mutation process is activated.

Homing Properties of B Cells to Primary and Secondary Follicles

Naive, virgin B cells arise from stem cells in the bone marrow throughout life. The peripheral B cell pool is composed of both a long-lived population that survives 1-4 weeks and a short-lived population that survives only 1-3 days.[16,17] IgM^+IgD^+ virgin B cells enter the spleen through the marginal zone by penetrating the marginal zone sinus and migrating along the outer zone of the PALS and then into the primary follicles. It is in the outer PALS of the spleen or lymph nodes that approximately 10% of the naive B cells are recruited into the long-lived pool and their migration into primary follicles is required for long-term B cell survival in the preimmune repertoire.[18] This selection process acts through sIg but does not involve isotype switching or somatic hypermutation of V genes.[19,20]

The cell-signal pathway to recruit naive B cells into the long-lived pool includes the BCR, which can signal either positively or negatively,[18,21,22] and the tyrosine phosphatase CD45.[18,23] Insight into directed migration of naive B cells into primary and then into secondary follicles has come from transgenic and targeted gene knockout mice, as detailed in chapter 1. The Burkitt's lymphoma receptor 1 (BLR1), a G protein-coupled putative chemokine receptor that is expressed on mature recirculating B cells but not newly produced B cells, appears to be a receptor for B cell homing to follicles in the spleen and Peyer's patches.[24] $BLR1^{-/-}$ mice lack inguinal lymph nodes and have few Peyer's patches and, after immunization, fail to develop germinal centers in the spleen in spite of generating PNA-binding B cells. Recirculating IgD^+ B cells fail to organize into discrete follicles, but accumulate at the periphery of the T zone as a thin band inside the marginal sinus. $TNF\alpha^{-/-}$ and $TNFRI^{-/-}$ also display a phenotype similar to $BLR1^{-/-}$ mice. Recently, a B cell homing or attracting CXC chemokine that functions as a BLR-1 ligand has been identified in lymphoid follicles.[24a,24b]

It is in the PALS that antigen-specific B cells encounter T cell help and are activated to proliferate. The activated B cell progresses into one of two pathways-differentiation to either become a plasma cell within the PALS, which may be accompanied by isotype switching, or initiation of a germinal center reaction in a primary follicle. The decision to proceed toward one or the other pathway and their relative degree of polarization is dictated by the degree and type of T cell help and the maturation state of the B cell—naive vs. memory.[25,26] CD40-CD40 ligand[27,28] or OX40 ligand-OX40[29,30] interactions between B and T cells are thought to promote germinal center precursor or plasma cell development respectively. Germinal center formation is inhibited by interference with CD40-CD40 ligand interactions.[27,28,31-35] After primary immunization, OX40 expressing T cells are detected in the PALS in proximity to antigen-specific B cell blasts, and inhibition of OX40-OX40 ligand interaction suppresses primary antibody responses.[30]

Germinal Center B Cell Subpopulations and Phenotypes (Tables 3.1 and 3.2)

Identification of surface markers expressed on germinal center cells or their subpopulations has greatly facilitated the studies of germinal centers in the last decade. Among the markers, a frequently used marker is peanut agglutinin (PNA), a lectin derived from the peanut plant *Arachis hypogaea* that recognizes terminal galactose residues. The antigen(s) that PNA binds has not been elucidated although PNA may bind CD45.[36] PNA does not identify all germinal center cells since there are PNA⁻ B cells in germinal centers of human or mice, and human germinal center T cells are not PNA⁺. PNA can be used to identify germinal centers in both human and mice,[37] but PNA does not identify germinal centers in all animals.

CD38

CD38 is another commonly used marker to identify human germinal center B cells and it is highly expressed on both human germinal center B cells and plasma cells. Human CD38 is a single-chain 45 kDa glycoprotein with ADP-ribosyl cyclase and hydrolase activity.[38,39] Crosslinking of CD38 with antibodies prevents apoptosis of human germinal center B cells but fails to induce proliferation.[40] A ligand for human CD38, designated Moon-1, has been identified on endothelial cells and at lower levels on naive T cells and monocytes.[41] Unlike human CD38, murine CD38 is expressed on follicular B cells but is downregulated on murine germinal center B cells located within the Peyer's patches.[42] In the spleens of immunized but not control mice, CD38$^{dim/-}$B220$^+$ germinal center B cells can be found. Also, mature murine plasma cells isolated from in vitro cultures fail to express CD38.

Table 3.1. Phenotypes of B cell subsets in germinal centers

GC Zone:	Follicular Mantle	Dark Zone	Basal Light Zone	Apical Light Zone	Apical Light Zone
Cell:	"Naive"	Centroblast	Centrocyte	Memory	Plasma Cell
Immunoglobulin					
sIgD	+	–	–	–	–
sIgM	+	–	+	+/–	–
sIgG	–	–	+	+	–
cIg	–	–	–	–	+
Activation/Proliferation/Differentiation					
CD10 (CALLA)	–	+	+	–	–
CD23 (FcεRII)	–/+	–	–	–	
CD38	–	+++	+	–	++
CD39	+	–	–	+	?
CD44 (pgp-1)	+	–	–/lo/+	+	+
CD71 (transferrin receptor)	–/+	+	+	–	–
CD77 (Gb3)	–	++	+	–	–
PNA binding site	–	+++	+	+	?
Ki67	–	+++	+	–	–
Bcl-6	–	+	+	–	–
Apoptosis/Selection					
Bcl-2	+	–	–	+	+
Fas/CD95	–	+	+	–	
p53	–	+	++	–	
c-Myc	–	++++	+	+	
Bax	–	++	+++	+	
Adhesion					
LFA-1 (CD18/11a)		+	+		
LFA-3		+	+		
L-Selectin	+	–	–	+/–	
Other					
CD5	+	–	–	–	
CD21(CR2)	+	–	–	–	–
CD40	+	+	+	+?	+?
CD70		–	+/–	–	
CD75	lo	+	+		
CD80 (B7-1)	–	++	+	+++	
CD86 (B7-2)	–	–	++	+++	
RAG-1/RAG-2	–	+	+	–	–
Telomerase	–	+++	+	–	–

Table 3.2. Germinal center cell receptor-ligand interactions

B Cell	T Cell	Follicular Dendritic Cell
sIg	Id-specific TCR	immune complexes
sIgD	IgD-R	
CD40/CD40-L	CD40-L	
Ia	TCR/CD4	
LFA-1 ($\alpha_L\beta_2$)	ICAM-1 (CD54)	ICAM-1 (CD54)
VLA-4 ($\alpha_4\beta_1$)		VCAM-1
c-Met tyrosine kinase receptor for HGF/SF		HGF/SF receptor
CD21		CD23, C3 on immune complexes
CD80/CD86	CD28/CTLA-4	
Fas/FasL	FasL	
CD19/CD77 (?)		CD19/CD77
CD28 plasmablast		GCDC-CD80/86
CD38		Moon-1 (endothelial cells)

CD77 (Gb3, BLA,CTH)

Globotriaosylceramide (CD77) is a neutral glycolipid expressed in germinal centers where it is most prominent on centroblasts in the dark zone.[43-45] CD77 functions as a receptor for the verotoxin (Shiga-like toxins) family of *E. coli* or shigella toxins.[46] It has been suggested that the extracellular domain of CD19 shares homology with verotoxins and may function with CD77 as a ligand-receptor pair, as evidenced by co-capping of CD19 and CD77 on the B cell surface.[47] Human tonsillar CD77[+] B lymphocytes express the B cell antigens CD19, CD20, CD21, CD22 and CD40, as well as CD38, and the adhesion molecules LFA-1, LFA-3 and CD44. They are positive for sIgM and negative for sIgD and for the classical activation antigens: CD23, CD25 (the IL-2 receptor α chain), and CD71 (the transferrin receptor). Only about one-half of the PNA[+] or CD38[+] subpopulation are CD77[+]. The morphology of CD77[+] B lymphocytes reveals this subset is undergoing apoptosis.[43] CD77 may be also expressed on germinal center B cells from mice. Globoside is another neutral glycolipid that is expressed on human germinal center B cells[48] but, in contrast to CD77, it may not be expressed on Burkitt's lymphoma cell lines.[48-50]

In addition to the increased expression of a cell surface marker, germinal center B cells can be identified by the decreased expression of markers. A classic marker of this type is IgD, as it is known that most (but not all) germinal center B cells express low levels of IgD on the surface. Other examples are CD24 and CD39. CD39[51] is an activation antigen that tends to be expressed later than CD25 and CD71, but is reduced in expression in germinal center B cells.

Subpopulations

Using a series of cell markers, several groups have distinguished subsets of B cells in murine lymph nodes and human tonsils. Based on the surface markers sIgD, CD38, and CD77 expressed on human tonsil B cells, naive B cells—sIgD$^+$CD38$^-$CD77$^-$, centroblasts—sIgD$^-$CD38$^+$CD77$^+$, and centrocytes—sIgD$^-$CD38$^+$CD77$^-$, have been isolated and characterized. Liu and colleagues identified seven subpopulations of B-cell subsets in human tonsils based on the markers sIgD, CD38, CD23, and CD77.[52] They separated "naive" IgD$^+$CD38$^-$CD23$^-$ B cells (Bm1 subset) from a CD23$^+$ (Bm2) subset, and suggested that the latter represented activated B cells. Whether the CD23$^+$ cell represents a ligand selected B cell, as discussed above, however, has not been determined. The sIgD$^-$CD38$^+$ B cells were divided into a CD77$^+$ (Bm3) or centroblast and a CD77$^-$ (Bm4) or centrocyte subset. The sIgD$^-$CD38 (Bm5) cells were designated as memory B cells.[53]

Rare subsets of B cells have also been detected in the germinal centers of human tonsils. An sIgD positive germinal center B cell subset has been detected by several investigators.[54,55] We have demonstrated an sIgD$^+$ PNA$^+$ cell subset that also expressed the activation marker CD23 and described its location in the dark zone in germinal centers in human tonsils. Others have characterized this subpopulation to express the germinal center markers: CD10, CD38, CD75, CD77 and CD95/Fas, and demonstrated their tendency to undergo spontaneous apoptosis in vitro and the failure of crosslinking of their antigen receptor to stimulate DNA synthesis. Following their stimulation in vitro, these cells predominantly secreted IgM and lacked surface expression of secondary isotypes. IgD$^+$ CD38$^+$ B cells were located in two distinct germinal center structures: either scattered cells within a typical germinal center or as a rare atypical germinal center with homogeneous IgD$^+$ B cells.[55] Liu and colleagues designated an sIgM$^+$sIgD$^+$CD38$^+$ subset (Bm2') as a germinal center founder B cell that had begun the transition to developing a germinal center phenotype with a tendency for apoptosis,[13] as detailed above.

An sIgD$^+$sIgM-CD38$^+$ subset (Bm3δ/4δ) that expresses highly mutated immunoglobulins has been isolated and characterized by Liu and colleagues.[56] It was suggested that through homologous recombination these cells have lost the ability to undergo isotype switching as evidenced by deletion of the C and S locus from V gene rearrangements. The B cells in this subset were commonly λ light chain positive, which was thought, but not proven, to arise from receptor editing in the germinal center with secondary light chain rearrangements at the light chain locus. Recent evidence suggests that receptor editing may in fact occur in the germinal center[57,58] through reactivation of RAG-1 and RAG-2.[59,60]

There are significant changes in the expression of Ig genes among the B cell subpopulations in germinal centers. The centroblasts in the dark zone downregulate the sIg that is expressed on follicular mantle zone cells with most cells expressing an sIgD^{lo} sIgM$^-$ phenotype. In the light zone, sIg is reexpressed. The isotype expressed in human tonsil is mainly IgG with smaller amounts of IgA and IgM. IgA predominates in Peyer's patches and mesenteric lymph nodes. Isotype switching occurs both in and outside of the germinal center. Isotype switching occurs in the PALS in the absence of somatic hypermutation after cognate T-B cell interactions.[25] Based on the isolation of Ig sterile transcripts and 5'S-S3' switch circles,[61] the isotype switching that occurs in the germinal center is thought to be ongoing in centrocytes in the light zone of human tonsils and appears to be occurring after somatic hypermutation has ensued. It is also in the light zone that T cells express CD40L[62] and are capable of secreting IL-4 and IL-10,[63] which may be directing isotype switching events. The dissociation of isotype switching from somatic hypermutation may prevent the generation of high affinity IgG autoantibodies prior to a stage when apoptosis is ongoing. On the other hand, the finding that CD44$^-$ murine centroblasts have undergone extensive isotype switching suggests that switching may occur in the dark zone.[64]

Surface Ig appears to play an important role in the processing and presentation of antigen by centrocytes to antigen-specific T cells in the light zone. The recruited T cells act to rescue B cells from apoptosis and stimulate B cell development into a memory B cell pathway. A role for antigen capture and presentation by B cells in the light zone is suggested by the decreased IgG1 and IgE antibody responses, affinity maturation, and the number of isotype-specific memory B cells in gene-targeted knockout mice with deletion of the cytoplasmic domain of γ1 or ε heavy chains, which is required for internalization of antigen and subsequent processing.[65-67]

Other interesting markers associated with germinal center B cell subpopulations are B7-1 and B7-2 costimulation molecules. In the mouse, B7-1 is expressed more on centroblasts and B7-2 is expressed more on centrocytes.[68,69] Interference with B7-2 interaction with CD28/CTLA4 during the late phase (day 6-10) of a primary antibody response interferes with somatic hypermutation of Ig genes and humoral memory responses.[31]

Recapitulation of Expression of Markers and Genes Expressed in Early B Cell Development

It has been suggested[59,70] that germinal center B cells demonstrate a recapitulation of gene expression and cell markers that are expressed on developing B cells in the bone marrow. Cell markers that are expressed include human CD38 and the heat stable antigen CD24. Cell proteins reexpressed include the Ig gene rearrangement mediating enzymes RAG-1 and RAG-2.[59,60,71] RAG expression is induced in germinal centers by immunization,[59,60] increases in time after immunization with development of more mature germinal center,[59] and is most prominent in apoptotic tingible bodies.[71] The ability of RAG expression to mediate V(D)J rearrangement to receptor-edit the antibody expressed by a B cell that encodes a self-reactive antibody is suggested by recent experiments demonstrating ongoing secondary rearrangements in germinal center B cells.[57,58,72]

Another enzyme that is reexpressed in germinal centers is telomerase.[73,74] The ribonucleoprotein telomerase, which maintains telomere length, is up-regulated as B cells differentiate from IgD$^+$CD38$^-$ naive B cells to become CD38$^+$IgD$^-$ B cells and is then downregulated as cells develop to become IgD$^-$CD38$^-$ memory B cells.[73,74] Telomere length is thought to serve as a mitotic clock for cell senescence, and therefore preservation of the length of the telomere in the highly replicative germinal center centroblast may act to maintain the longevity of the memory B cell.

Neutral endopeptidase (CD10) and carboxypeptidase M are expressed on B cell precursors in the bone marrow.[75] Human germinal center B cells express increased amounts of carboxyl-peptidase M.[75] Although CD10 expression is increased on human germinal center B cells,[44,76] CD10 expression is decreased in mouse germinal center.[77] Germinal center T cells are highly sensitive to apoptosis and express Nur77, a gene that is associated with the T-cell receptor and mediates apoptosis in thymocytes.[78,79]

B Cell Markers and Genes that are Associated with Cell Adhesion

LFA-1/ICAM-1; VLA-4/VCAM-1

Within secondary follicles, PNA$^+$ dividing centroblasts and FDC make intimate contact through complementary surface antigens. The integrin leukocyte function-associated antigen-1 (LFA-1, CD11a/CD18) on B cells and intercellular adhesion molecule 1 (ICAM-1, CD54) on FDC and the integrin very late activation antigen-4 (VLA-4, CD69d,α4/β1) on B cells and vascular adhesion molecule-1 (VCAM-1) on FDC are two of these interacting receptor-ligand pairs. VLA-4 (4/1) - VCAM-1 interactions can rescue germinal center B cells from apoptosis.[80-82]

CD44 (HK23)[44]

A glycoprotein with homology to cartilage link and proteoglycan proteins that binds to hyaluronates, is expressed on resting sIgD$^+$sIgM$^+$ cells and is then downregulated as CD38 is upregulated on centroblasts,[64,83-85] with the majority of CD38$^+$sIgM$^+$ or sIg$^-$ centroblasts being CD44$^-$. As centrocytes develop, CD44 is reexpressed on the cell surface, although some centrocytes remain CD44$^-$. CD77$^+$ B cells tend to be CD44$^-$.[44,64] Cross linking of CD44 promotes adhesion of B cells to FDC.[64]

CD23 (Blast-2)

CD23 is an activation antigen that is expressed variably in PNA$^+$ germinal center B cells in only some human tonsils.[44,54] FDC also express CD23, a low-affinity receptor for IgE and for CD21(CR2/Epstein-Barr virus receptor). Those FDC furthest from centroblasts express higher levels of CD23.[86] Soluble CD23, especially in the presence of interleukin-1α and monoclonal antibodies to CD21, has been shown to potentiate the survival of germinal center B cells in vitro and promote their differentiation toward a plasmacytoid morphology.[86,87] Also, fragments of C3 on FDC interact with CD21 on germinal center B cells.

HGF/SF

The c-Met-hepatocyte growth factor/scatter factor (HGF/SF) pathway also may be involved in germinal center B cell adhesion to FDC.[88] The c-Met tyrosine kinase receptor for HGF/SF is preferentially expressed on $CD38^+CD77^+$ centroblasts and is upregulated by ligation of CD40 on the B cell surface. Activation with HGF/SF, which is produced by FDC, stimulates enhanced VLA-4 mediated adhesion of B cells to VCAM-1 and fibronectin.

Connexin43

Connexin43 gap junctions are present in the light zone of the germinal center and appear to couple FDC to FDC and germinal center B cells to FDC.[89]

CD73

CD73 (ecto-5'-nucleotidase, lymphocyte-vascular adhesion protein-2) on the FDC can mediate adhesion to B cells based on in vitro inhibition of their aggregation with monoclonal antibodies and the demonstration that CD73 is expressed in vivo on FDC.[90]

L-Selectin

L-selectin, recognized by antibody Leu8 to human or mel-14 antibody to mouse L-selectin, is not expressed in the germinal center.[91]

B Cell Markers and Genes that Are Associated with Apoptosis and Selection

Centroblasts quickly die by apoptosis in vitro, in contrast to centrocytes, in which apoptosis is delayed.[82] Germinal center B cells express c-Myc, p53, Bax, and Fas, but not Bcl-2.[92] The apoptosis-inducing genes—c-Myc, p53, Bax, and Fas—and the survival gene Bcl-2 are differentially expressed during B cell differentiation in secondary lymphoid organs. c-Myc expression is higher in centroblasts than centrocytes, which is associated with the greater degree of proliferation in those cells, but is still prominent in centrocytes, which suggests that their mechanism of apoptosis must be dictated by other genes. The increased levels of c-Myc may contribute to both the high degree of centroblast proliferation and to priming for apoptosis.[93,94] Transfection of c-Myc into EBV-transformed cells generates a centroblast-like cell that expresses CD10 and CD38 and develops susceptibility to spontaneous apoptosis.[94] Bax and p53 expression are higher in centrocytes than in centroblasts. Whether the increased expression of p53 as germinal center B cells transit to a centrocyte stage reflects an accumulation of damaged DNA with double strand breaks from the mutation process is unknown.

Positive selection is required to select high-affinity mutated B cells and negative selection is required to delete or neglect self-reactive B cells that are generated by somatic hypermutation. Positive selection of germinal center B cells is thought to be mediated by interaction of: the BCR with antigen; CD40 on the B cell with CD40 ligand (CD40L)[27,28,95,96] on activated germinal center T cells; and other costimulatory receptor-ligand interactions between centrocytes and T cells or FDC. CD40L is expressed only transiently on activated T cells because of induction of endocytosis after engaging CD40. CD40L-CD40 interactions in vitro stimulate germinal center B cell survival, proliferation, and inhibition of differentiation to become an antibody-secreting cell. Generation of B cell memory is inhibited by soluble CD40 or anti-CD40L Ab by interfering with this interaction.[31,97] The mechanism of rescue from apoptosis by interactions of the B cell with FDC is not completely understood, but it has been shown that FDC can irreversibly block the endonuclease that mediates apoptosis.[98]

Germinal center B cells express high levels of functional CD95 (Fas).[99-101] Ligation of Fas on B cells by FasL-expressing T cells will lead to B-cell death and is prevented by antigen cross linking of the BCR.[102] This may be a mechanism by which high affinity binding B cells are distinguished from low affinity and autoreactive B cells. Germinal center B cells that are rescued from apoptosis by CD40 ligation or attachment to FDC become resistant to Fas-Fas ligand induced apoptosis.[99] Signaling via CD40-CD40L abrogates induction of apoptosis of germinal center B cells via Fas over only a short time period, with induction of apoptosis occurring after 48 hours of culture in vitro. CD40L signals alone, in the absence of BCR crosslinking, may lead to enhanced killing by Fas ligation[102-105] or by subsequent crosslinking of the B cell receptor.[106,107] Fas ligand is also induced on activated B cells[108] and potentially could negatively select Fas-expressing germinal center B cells with activation of apoptosis being primed by preactivation via CD40 or inhibited by preactivation by signaling through the BCR.

B Cell Markers and Genes that Are Associated with Proliferation and Differentiation

In addition to the genes described above that contribute to cell growth and death in germinal centers, there are cell markers and transcription factors expressed in centroblasts and centrocytes that affect proliferation, and the decision for a germinal center B cell to develop into a memory B cell or a terminally differentiated plasma cell. No surface markers have been defined that are unique to memory B cells. Typically they have been defined as sIgD⁻CD38⁻ B cells that have a lower threshold for activation, the capacity to rapidly upregulate costimulatory molecules (e.g., CD80,CD86),[53] and the ability to persist in the host. Memory B cells are found in the circulation, the splenic marginal zones, as residual cells in follicles, or in mucosal epithelium, such as under the dome of Peyer's patches or in crypts of the palantine tonsil.[53,109-113] Antigen-specific follicular and memory B cells can continue to proliferate for months after the initiation of a T-cell dependent antibody response and not uncommonly express both sIg and cIg, suggesting that the germinal center is a site of late antibody formation and maintenance of memory.[114] It has been suggested that in contrast to naive B cells, reactivation of memory B cells leads to a greater tendency to generate a plasma cell response than a memory B cell response with germinal center-type reactions.[4,115,116] In vitro ligation of CD40 on the B cell tends to promote differentiation toward the memory cell pathway and inhibition of differentiation toward plasma cell development, suggesting that centrocyte interaction with CD40L-expressing T cells in the light zone will activate this pathway. Administration of anti-CD40L antibody can abrogate an established germinal center reaction.[31] T cell help can direct development of plasma cells based on OX40-OX40 ligand T cell-B cell interactions.[29,30] Cytokines that stimulate differentiation to a plasma cell stage include: IL-2, IL-6, IL-10, and recombinant soluble CD23, which stimulates B cells through CD21[87] in the presence of IL-1.[86]

Pax5/BSAP

The DNA-binding transcription factor B cell-specific activator protein (BSAP), a product of the Pax 5 gene, is expressed in all developmental stages of B cells up to the terminally differentiated plasma cell; it stimulates B cell proliferation and suppresses B cell differentiation.[117-122] BSAP upregulates B cell proliferation, is required for germline Ig gene transcription prior to isotype switching and inhibits differentiation. The differentiation of B cells to plasma cells is associated with downregulation of BSAP. Binding sites for BSAP are present in the upstream region of several genes, including CD19, Blk, and λ5. Alternatively spliced isoforms of Pax-5 may be involved in gene regulation.[123]

Blimp-1

One transcription factor that appears to be important for B cell differentiation to plasma cells is Blimp-1 or its human homologue PRD1-BF1, a zinc finger-containing transcription factor that is rapidly induced during the differentiation of B lymphocytes to immunoglobulin secreting cells.[124,125] Transfection of Blimp-1 into B lymphoma lines leads to induction of J chain transcripts, immunoglobulin secretion, and up-regulation of Syndecan-1.[126] Blimp-1 appears to act by repression of c-Myc transcription.[127] A physiologic role for Blimp-1 is not defined at this time and it is unclear whether c-Myc repression alone is sufficient to induce B cell differentiation or whether Blimp-1 responsive genes are part of a general cascade for regulating differentiation to plasma cells.

Ki67

Ki67 is a cell cycle-associated nuclear antigen that is highly expressed in centroblasts. Human germinal center B cells express increased amounts of CD71 (transferrin receptor).

Other Unique B Cell Markers and Gene Expression

Bcl-6

The Bcl-6 gene encodes a transcriptional repressor that has six C-terminal zinc finger motifs, which are homologous to members of the Kruppel subfamily of zinc finger proteins, and an N-terminal ZiN/POZ (zinc finger N-terminal/ POx zinc finger) domain, which is observed in other mammalian zinc finger proteins. Expression of Bcl-6 is most prominent in germinal center B cells where it is expressed in both centroblasts and centrocytes.[128,129] Follicular mantle zone B cells and plasma cells fail to express the Bcl-6 protein. As detailed in chapter 5, the Bcl-6 gene is involved in chromosomal translocations in non-Hodgkin's lymphomas at a frequency of 30-40% of diffuse large-cell lymphoma (DLCL), 5-14% of follicular lymphomas, and 20% of AIDS-associated DLCL. Bcl-6 expression is regulated in lymphocytes during mitogenic stimulation with a 5- to 35-fold decrease in Bcl-6 mRNA levels from resting levels after stimulation.[130] In contrast, replicating human germinal center B cells express Bcl-6 mRNA at levels comparable to the levels in resting B cells and protein levels 3- to 34-fold higher than in resting B cells, suggesting that the germinal center reaction provides unique activation signals to maintain high-level Bcl-6 expression. The finding that Bcl-6 is specifically expressed in germinal center B cells and is abnormally expressed in tumors of germinal center B cells suggests that Bcl-6 plays a role in germinal center development or function. Bcl-6 gene targeted knock-out mice lack germinal centers, have normal primary follicles, and display defective antibody responses to TD but not to TI type 2 (e.g., TNP-Ficoll) antigens.[131,132] The Bcl-6 DNA recognition motif resembles sites bound by the STAT transcription factors and can block Stat6-dependent activation of reporter genes, including transcription of IL-4.[131] This suggests that Bcl-6 may modify the effects of cytokines that act through STAT factors.

BL44

The *BL44* gene encodes for a germinal center kinase (GCK), which can activate SAPK (stress-activated protein kinase), but although this kinase is differentially expressed in germinal center compared to follicular mantle B cells, the kinase is expressed in several other tissues.[133] Both GCK and SAPK are inducible with TNF-α.

M17

M17[134] is a cytoplasmic protein with homology to lipid binding amphipathic helices that may play a role in apoptosis.

MHC Class II

MHC class II antigens are expressed at increased levels by germinal center B cells.[135] A marker that has been used to identify cells that are activated in germinal centers is the MHC class II I-E, which is, paradoxically, expressed if it is under control of its own promoter with a deletion in its 5' flanking region.[21,136,137]

8-Oxoguanine DNA Glycosylase

8-Oxoguanine DNA Glycosylase (MutM), a DNA repair enzyme that repairs oxidative damage to guanine on DNA, is expressed in centroblasts in the dark zone of human germinal centers.[138] The functional role for this enzyme in germinal centers is unknown at this time.

SHP-1

SHP-1, a phosphotyrosine phosphatase that negatively regulates B cell activation, is downregulated in germinal center B cells,[139] suggesting that germinal center B cells may have a decreased threshold for activation.

Other Markers

Germinal center B cells may express EAA-B, which is an 80-90 kDa molecule[140] and is expressed on activated endothelial cells. CD75 is an antigen recognized by LN-1[141] that is expressed in increased amounts on germinal center B cells compared to other B cells. The antigen is 2,6-linked sialic acid residue and may be a ligand for CD22β.[142]

Germinal Center T Cells

The critical importance of T cells to germinal center formation was quickly appreciated following the discovery of athymic nude mice. It was observed that these mice have markedly reduced numbers of germinal centers[143,144] and that they can readily develop germinal centers following the transfer of normal T cells to them.[143,145] The importance of T cells in germinal center formation was further appreciated when it was found that germinal centers could be reliably induced with T cell-dependent antigens but not with many (though not all[144,146]) T cell-independent antigens.[144,147]

When T cells could be identified in situ, it was found that some cells in germinal centers are T cells (Fig. 3.1).[148] The T cells are not randomly located in the germinal centers but are preferentially located in the light zone of germinal centers, and T cells can be up to 10-20% of the cells in the light zone of the germinal centers.[149] The light zone has been divided into two zones named apical and basal light zones based on the expression of CD23.[150] In many tonsils, the germinal center T cells are more common in the apical light zone near the mantle zone than in the basal light zone.[151]

Examination of germinal center T cell phenotypes showed that a great majority of the germinal center T cells are CD4$^+$, CD8$^-$ in mice[152] as well as in humans.[149,153] Consistent with this observation, mice with absence of MHC class II genes have no germinal centers.[154] Adoptive transfer studies showed CD4$^+$ T cells to be much more efficient than CD8$^+$ cells in allowing athymic mice to form germinal centers.[145] In addition, germinal center T cells are predominantly

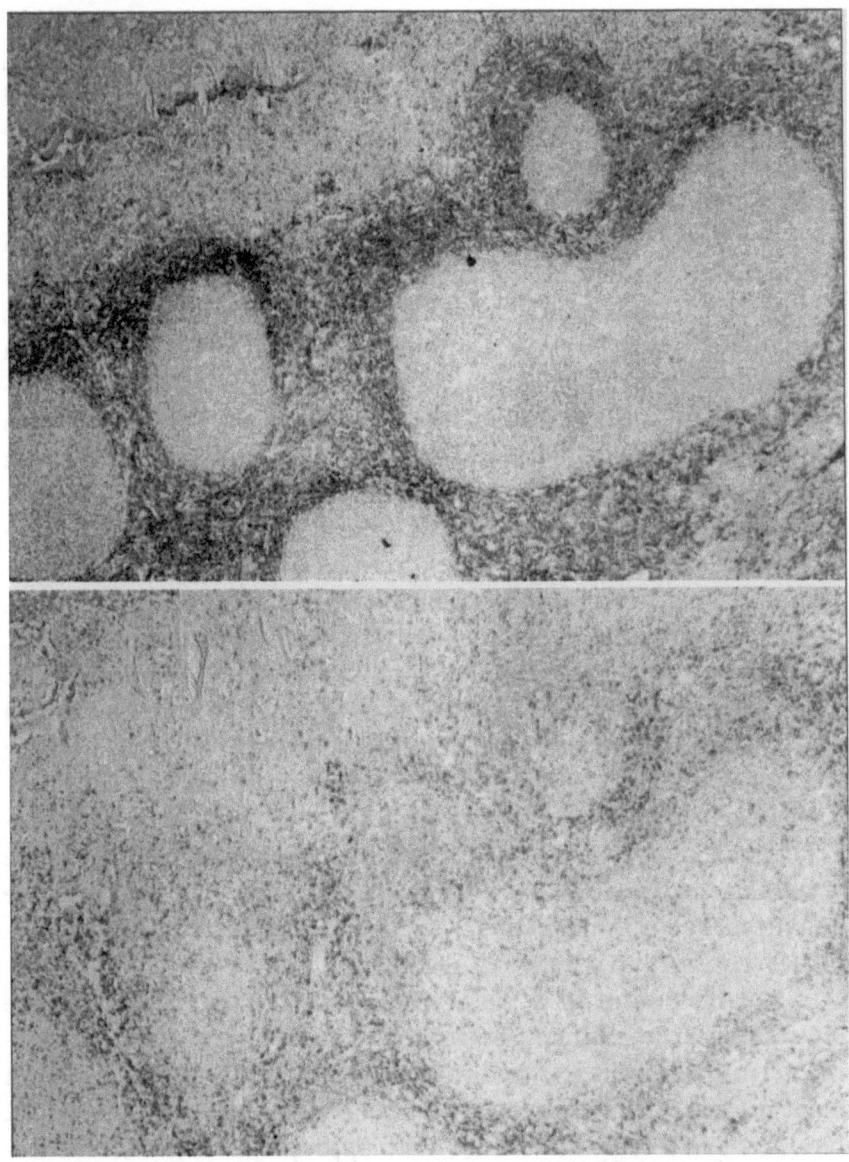

Fig. 3.1. Serial sections of a tonsil were immunohistochemically stained (brick-red) for L-selectin (top panel) or for CD3 (bottom panel) and counterstained (green) with methyl green pyronin. In the top panel, germinal centers appear as unstained areas circumscribed by L-selectin[+] cells. In the U shaped germinal center, dark and light zones appear as the solid green (bottom) or light green (top) stained areas. In the bottom panel, CD3[+] cells are in the light zone of the germinal centers. In the bottom panel, mantle zones containing the resting B cells appear as intensely green areas on top of germinal centers. See color figure, page 239.

CD45R0[+] and are a distinct memory-type subset of CD4[+] T cells.[151,155] Nevertheless, CD8[+] cells are present in germinal centers[149,152] and a high (up to 10%) number of CD8[+] cells can be found in occasional germinal centers of human tonsils.[149] Also, mice lacking the CD4 gene are capable of forming germinal centers.[156] Therefore, CD4[+] cells are not absolutely needed to form germinal centers.

Germinal center T cells display various markers associated with cell activation. About half of germinal center T cells in human tonsils express CD40 ligand,[157] a cofactor for B cell stimulation that is upregulated with T cell activation. Expression of CD40 ligand on mouse germinal center T cells was not demonstrated.[158] Mouse germinal center T cells express CD28 and CTLA-4.[69] These molecules are critical for the formation of germinal centers, since neutralizing their activity abrogates germinal center development.[159,160] The germinal center T cells expressing CD57 also express an early activation marker, CD69, but do not express ferritin receptor or IL-2 receptor.[151] The diversity in germinal center T cell phenotypes does indicate the presence of heterogeneity among germinal center T cells.

Most of the germinal center T cells expressing $V\alpha$ and $V\beta$ TCR. However, germinal center T cells are not limited to express $V\alpha$ and $V\beta$ TCR families, as germinal centers can be readily found in mice lacking TCR $V\alpha$.[156,161] Interestingly, the majority of the germinal center T cells in the TCR $V\alpha^{-/-}$ mice express $V\beta$ without expressing pre-$T\alpha$ chain[162] and $V\gamma$ and $V\delta$, although the majority of the cells in the T cell zone express $V\gamma$ and $V\delta$.[156] Mice lacking a TCR $V\beta$ chain are able to make germinal centers, but the observed frequency of germinal centers is strikingly reduced.[156,163]

Further studies of TCR showed that germinal center T cells are specific for the immunizing antigen. This was demonstrated by studying germinal centers induced with pigeon cytochrome c and myelin basic protein. These antigens stimulate T cells expressing select TCR families which can be identified with appropriate antibodies. Cytochrome c mainly stimulates T cells expressing $V\alpha11$ and $V\beta3$ families, and myelin basic protein primarily stimulates the T cells expressing the $V\beta8$ TCR family. With germinal centers induced with cytochrome c, the T cells in germinal centers primarily express the $V\alpha11$ TCR family, but not the T cells expressing other TCR families (Figs. 3.2 and 3.3).[164] Also, germinal centers induced with myelin basic protein preferentially attract T cells expressing $V\beta8$ TCR families (Fig. 3.3).

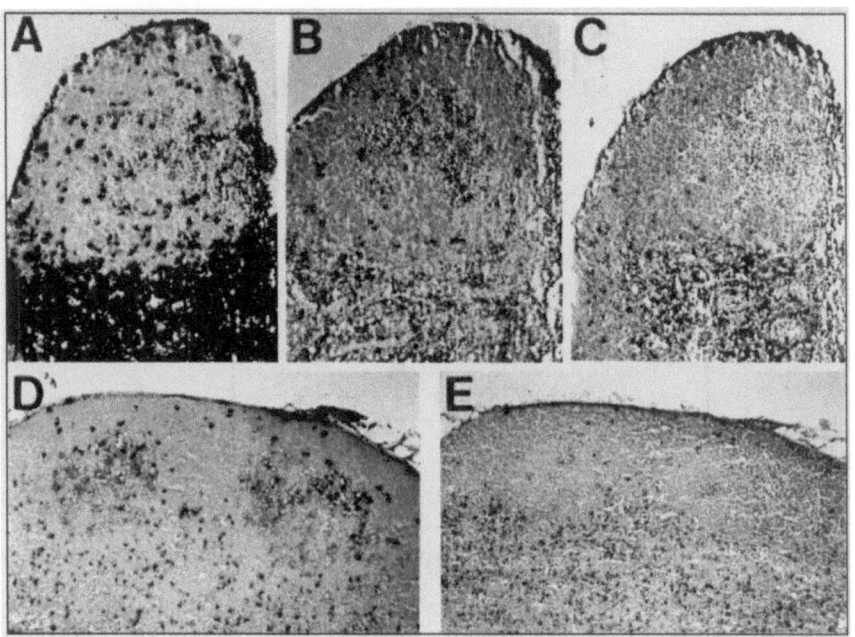

Fig. 3.2. T cells in germinal centers induced with cytochrome c. Cells in popliteal lymph node section were stained black with anti-Thy1.2 antibody (A) or with anti-Vα11 antibody (B and D), or with anti-Vβ8 antibody (C and E). Germinal centers are visualized as pink areas in the lymph node, which is counterstained green with methyl green. A, B, and C were adjacent sections from one lymph node, and D and E were from another. The popliteal lymph node was harvested 9 days (A to C) or 7 days (D and E) after immunization. The following magnifications were used. A to C x100, D and E x40. Reprinted with permission from Fuller et al, J Immunol 1993; 151:4505-4512. Copyright 1993, The American Association of Immunologists. See color figure, page 240.

This finding was extended by Zheng et al[165] and Gulbranson-Judge et al[166] with the use of mice immunized with cytochrome c. They showed that the number of antigen-specific (i.e., $V\alpha11^-V\beta3^+$) T cells peaks in PALS on day 7 but quickly decreases. By day 14, the number of the antigen-specific T cells in the B cell follicles peaks and is more than that of the antigen-specific T cells in the PALS. More than half of the proliferating B cells in the B cell follicles are adjacent to the T cells.[167] Another confirmation was provided by showing that a large percent of T cells in germinal centers have recently proliferated since exposure to the antigen.[166,168] Lastly, this is confirmed with the use of adoptively transferring ovalbumin (OVA)-specific T cells from transgenic mice to congenic mice. It was shown that the adoptively transferred T cells migrate to the B cell follicles once activated with the antigen in the presence of an adjuvant.[169,170]

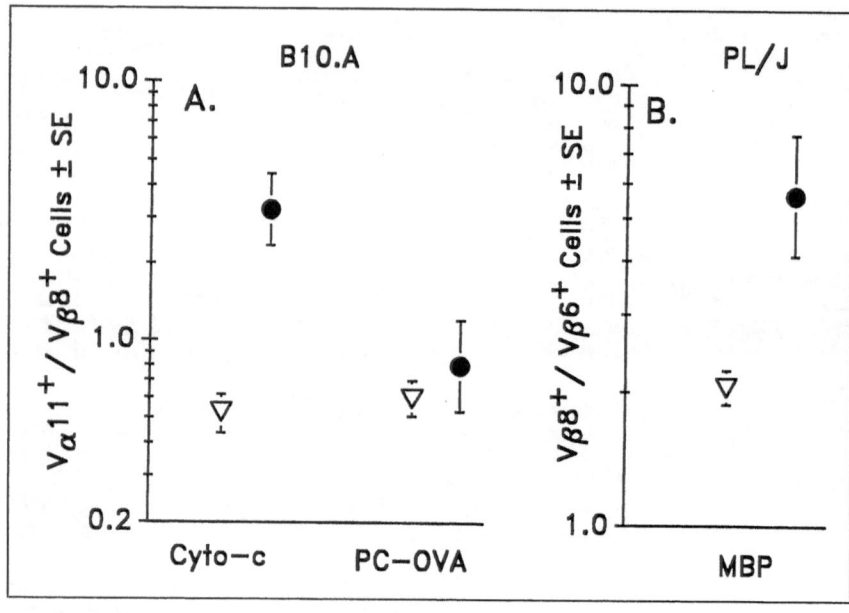

Fig. 3.3. Relative prevalence of T cells expressing a select TCR family in germinal center (solid circle) and paracortex (open triangle). (A) Prevalence ration of Vα11+ to Vα8+ T cells in lymph nodes of 11 B10.A mice stimulated with cytochrome c or PC-OVA. (B) Prevalence ratio of Vβ8+ to Vβ6+ T cells in lymph nodes of 12 PL/J mice stimulated with MBP. Data were analyzed with paired Student's t-test. Bars, SE. Reprinted with permission from Fuller et al, J Immunol 1993; 151:4505-4512. Copyright 1993, The American Association of Immunologists.

The group led by Kelsoe further extended the studies of TCR of germinal center T cells by determining the sequence of TCR mRNA of the T cells microdissected from germinal centers and showing that the TCR sequences are appropriate for the antigen.[165] They reported that the sequences show somatic mutations (in α but not in β chains) which may increase the avidity of germinal center T cells to the antigen.[171] Furthermore, they showed that the TCR mRNA sequences of the T cells in nearby germinal centers are identical, indicating that the T cells in two different germinal centers must have originated from one progeny T cell.

At the moment there are no phenotypic markers that can be used to purify all the germinal center T cells. Although most of the germinal center B cells express the binding sites for peanut agglutinin (PNA), and PNA is often used to isolate germinal center B cells, human germinal center T cells are PNA−.[151] Human germinal center T cells uniquely express CD57,[151,172] which is sulfated glucuronic acid[173] binding L- or E-selectin.[174] Although CD57 expression is used to purify human germinal center T cells, it is not expressed on all human germinal center T cells. Although Thy-1 expression has been used to histologically visualize the T cells in germinal center,[152,164] a recent study reported that germinal center T cells express a reduced level of Thy-1.[164a] The reduced expressed of Thy-1 may be useful in isolating germinal center T cells from mice.

In terms of function, germinal centers may be a site where T cells expressing desired TCR are selected, as they are for B cells. McHeyzer-Williams and Davis purified T cells specific for the immunizing antigen by flow cytometry from lymph nodes and showed that the TCR repertoire used by the antigen-specific T cells evolves quickly during the immunization. In the beginning, a diverse range of TCR was used, but two weeks after immunization the T cells expressing specific TCR combinations dominated.[175] When the TCR cDNA sequences of the T cells in the germinal centers were examined, they were diverse at the beginning of the immunization but converged to a few select TCR. In support of this evolution in the TCR repertoire in germinal centers, recent studies found that germinal center T cells do proliferate as well as they are prone to apoptosis[165] although mitotic germinal center T cells are not readily apparent.[167,176] Interestingly, 17% of germinal center T cells express Nur77,[165] which is associated with the apoptosis of thymocytes and T cell hybridomas.[78] The nature of the stimulations important for the proliferation and apoptosis of germinal center T cells is unknown. Recently described germinal center dendritic cells[177] may be important in stimulating the T cells in addition to the B cells in germinal centers.

Another potential function of germinal center T cells is to influence B cell maturation in germinal centers. For instance, germinal center T cells are likely to influence the isotype switching by B cells in germinal centers. B cells undergo isotype switching in germinal centers,[61] and human germinal center T cells are shown to express CD40 ligand, a factor necessary for isotype switching.[157] In addition, germinal center T cells express mRNA for IL-4,[63] which is well known to influence the isotype switching by B cells. In addition to isotype switching, the germinal center T cells may influence the germinal center B cell maturation into either memory cells or plasma cells in germinal centers. Cytokines and cofactors important for the maturation along these pathways are beginning to be identified.[30] CD40-CD40 ligand interaction has been reported to be important for memory B cell development.[115] OX40-OX40 ligand interaction has been reported to be important for plasma cell development.[30] Although some of these cofactors have been found expressed on germinal center T cells (e.g., CD40 ligand), expression of some of these cofactors (e.g., OX40) has not been fully studied. Additional studies in the future should better define the role of germinal center T cells in B cell differentiation.

References

1. MacLennan IC. Germinal centers. [Review] [139 refs]. Ann Rev Immunol 1994; 12:117-139.
2. Thorbecke GJ, Amin AR, Tsiagbe VK. Biology of germinal centers in lymphoid tissue. [Review] [103 refs]. FASEB Journal 1994; 8:832-840.
3. Tsiagbe VK, Inghirami G, Thorbecke GJ. The physiology of germinal centers. [Review] [432 refs]. Criti Rev Immunol 1996; 16:381-421.
4. Liu YJ, Zhang J, Lane PJ, Chan EY, MacLennan IC. Sites of specific B cell activation in primary and secondary responses to T cell-dependent and T cell-independent antigens [published erratum appears in Eur J Immunol 1992 Feb;22(2):615]. Eur J Immunol 1991; 21:2951-2962.
5. Vonderheide RH, Hunt SV. Comparison of IgD$^+$ and IgD$^-$ thoracic duct B lymphocytes as germinal center precursor cells in the rat. International Immunology 1991; 3:1273-1281.
6. Seijen HG, Bun JC, Wubbena AS, Lohlefink KG. The germinal center precursor cell is surface mu and delta positive. Adv Exp Med & Biol 1988; 237:233-237.
7. Vonderheide RH, Hunt SV. Surface IgD phenotype of rat germinal centre precursor cells. Adv Exp Med & Biol 1988; 237:239-243.
8. Amin AR, Swenson CD, Xue B, Ishida Y, Nair BG, Patel TB et al. Regulation of IgD receptor expression on murine T cells. II. Upregulation of IgD receptors is obtained after activation of various intracellular second-messenger systems; tyrosine kinase activity is required for the effect of IgD. Cell Immunol 1993; 152:422-439.
9. Roes J, Rajewsky K. Immunoglobulin D (IgD)-deficient mice reveal an auxiliary receptor function for IgD in antigen-mediated recruitment of B cells. J Exp Med 1993; 177:45-55.
10. Carsetti R, Kohler G, Lamers MC. A role for immunoglobulin D: Interference with tolerance induction. Eur J Immunol 1993; 23: 168-178.
11. Linton PJ, Lo D, Lai L, Thorbecke GJ, Klinman NR. Among naive precursor cell subpopulations only progenitors of memory B cells originate germinal centers. Eur J Immunol 1992; 22:1293-1297.
12. Kroese FG, Seijen HG, Nieuwenhuis P. The initiation of germinal centre reactivity. [Review] [24 refs]. Research in Immunology 1991; 142:249-252.

13. Lebecque S, de Bouteiller O, Arpin C, Banchereau J, Liu YJ. Germinal center founder cells display propensity for apoptosis before onset of somatic mutation. J Exp Med 1997; 185:563-571.

14. Matsumoto M, Lo SF, Carruthers CJ, Min J, Mariathasan S, Huang G et al. Affinity maturation without germinal centres in lymphotoxin-alpha-deficient mice. Nature 1996; 382:462-466.

15. Chu YW, Marin E, Fuleihan R, Ramesh N, Rosen FS, Geha RS et al. Somatic mutation of human immunoglobulin V genes in the X-linked HyperIgM syndrome. J Clin Invest 1995; 95:1389-1393.

16. MacLennan IC, Oldfield S, Liu YJ, Lane PJ. Regulation of B-cell populations. [Review] [73 refs]. Current Topics in Pathology 1989; 79:37-57.

17. MacLennan I, Chan E. The dynamic relationship between B-cell populations in adults [see comments]. [Review] [43 refs]. Immunol Today 1993; 14:29-34.

18. Goodnow CC, Cyster JG, Hartley SB, Bell SE, Cooke MP, Healy JI et al. Self-tolerance checkpoints in B lymphocyte development. [Review] [475 refs]. Advances in Immunology 1995; 59:279-368.

19. Chan EY, MacLennan IC. Only a small proportion of splenic B cells in adults are short-lived virgin cells. Eur J Immunol 1993; 23:357-363.

20. Lortan JE, Roobottom CA, Oldfield S, MacLennan IC. Newly produced virgin B cells migrate to secondary lymphoid organs but their capacity to enter follicles is restricted. Eur J Immunol 1987; 17:1311-1316.

21. Fehling HJ, Viville S, van EW, Benoist C, Mathis D. Fine-tuning of MHC class II gene expression in defined microenvironments. [Review] [29 refs]. Trends in Genetics 1989; 5:342-347.

22. Cyster JG, Hartley SB, Goodnow CC. Competition for follicular niches excludes self-reactive cells from the recirculating B-cell repertoire [see comments]. Nature 1994; 371:389-395.

23. Cyster JG, Healy JI, Kishihara K, Mak TW, Thomas ML, Goodnow CC. Regulation of B-lymphocyte negative and positive selection by tyrosine phosphatase CD45. Nature 1996; 381:325-328.

24. Forster R, Mattis AE, Kremmer E, Wolf E, Brem G, Lipp M. A putative chemokine receptor, BLR1, directs B cell migration to defined lymphoid organs and specific anatomic compartments of the spleen. Cell 1996; 87:1037-1047.

24a. Gunn MD, Ngo VN, Ansel KM, Ekland EH, Cyster JG, Williams LT. A B-cell-homing chemokine made in lymphoid follicles activates Burkitt's lymphoma receptor-1. Nature 1998; 391:799-803.

24b. Legler DF, Loetscher M, Roos RS et al. B cell-attracting chemokine 1, a human CXC chemokine expressed in lymphoid tissues, selectively attracts B lymphocytes via BLR1/CXCR5. J Exp Med 1998; 187:655-660.

25. Maclennan IM, Gulbransonjudge A, Toellner KM, Casamayorpalleja M, Chan E, Sze DY et al. The changing preference of T and B cells for partners as T-dependent antibody responses develop [Review]. Immunol Rev 1997; 156:53-66.

26. Liu YJ, Arpin C. Germinal center development [Review]. Immunol Rev 1997; 156:111-126.
27. Banchereau J, Bazan F, Blanchard D, Briere F, Galizzi JP, van Kooten et al. The CD40 antigen and its ligand. [Review] [233 refs]. Annual Review of Immunology 1994; 12:881-922.
28. Clark LB, Foy TM, Noelle RJ. CD40 and its ligand. [Review] [152 refs]. Advances in Immunology 1996; 63:43-78.
29. Stuber E, Neurath M, Calderhead D, Fell HP, Strober W. Crosslinking of OX40 ligand, a member of the TNF/NGF cytokine family, induces proliferation and differentiation in murine splenic B cells. Immunity 1995; 2:507-521.
30. Stuber E, Strober W. The T cell-B cell interaction via OX40-OX40L is necessary for the T cell-dependent humoral immune response. J Exp Med 1996; 183:979-989.
31. Han S, Hathcock K, Zheng B, Kepler TB, Hodes R, Kelsoe G. Cellular interaction in germinal centers. Roles of CD40 ligand and B7-2 in established germinal centers. J Immunol 1995; 155:556-567.
32. Kawabe T, Naka T, Yoshida K, Tanaka T, Fujiwara H, Suematsu S et al. The immune responses in CD40-deficient mice: Impaired immunoglobulin class switching and germinal center formation. Immunity 1994; 1:167-178.
33. Xu J, Foy TM, Laman JD, Elliott EA, Dunn JJ, Waldschmidt TJ et al. Mice deficient for the CD40 ligand [published erratum appears in Immunity 1994 Oct;1(7):following 613]. Immunity 1994; 1:423-431.
34. Castigli E, Alt FW, Davidson L, Bottaro A, Mizoguchi E, Bhan AK et al. CD40-deficient mice generated by recombination-activating gene-2-deficient blastocyst complementation. Proc Natl Acad Sci USA 1994; 91:12135-12139.
35. Renshaw BR, Fanslow WC, Armitage RJ, Campbell KA, Liggitt D, Wright B et al. Humoral immune responses in CD40 ligand-deficient mice. J Exp Med 1994; 180:1889-1900.
36. Thomas ML. The leukocyte common antigen family. [Review] [137 refs]. Ann Rev Immunol 1989; 7:339-369.
37. Rose ML, Birbeck MS, Wallis VJ, Forrester JA, Davies AJ. Peanut lectin binding properties of germinal centres of mouse lymphoid tissue. Nature 1980; 284:364-366.
38. Lund F, Solvason N, Grimaldi JC, Parkhouse RM, Howard M. Murine CD38: An immunoregulatory ectoenzyme. [Review] [46 refs]. Immunol Today 1995; 16:469-473.
39. Shubinsky G, Schlesinger M. The CD38 lymphocyte differentiation marker—new insight into its ectoenzymatic activity and its role as a signal transducer [Review]. Immunity 1997; 7:315-324.
40. Zupo S, Rugari E, Dono M, Taborelli G, Malavasi F, Ferrarini M. CD38 signaling by agonistic monoclonal antibody prevents apoptosis of human germinal center B cells. Eur J Immunol 1994; 24:1218-1222.
41. Deaglio S, Dianzani U, Horenstein AL, Fernandez JE, van KC, Bragardo M et al. Human CD38 ligand. A 120-KDA protein predominantly expressed on endothelial cells. J Immunol 1996; 156:727-734.

42. Oliver AM, Martin F, Kearney JF. Mouse CD38 is down-regulated on germinal center B cells and mature plasma cells. J Immunol 1997; 158:1108-1115.

43. Mangeney M, Richard Y, Coulaud D, Tursz T, Wiels J. CD77: An antigen of germinal center B cells entering apoptosis. Eur J Immunol 1991; 21:1131-1140.

44. Fyfe G, Cebra-Thomas JA, Mustain E, Davie JM, Alley CD, Nahm MH. Subpopulations of B lymphocytes in germinal centers. J Immunol 1987; 139:2187-2194.

45. Butch AW, Nahm MH. Functional properties of human germinal center B cells. Cell Immunol 1992; 140:331-344.

46. Waddell T, Cohen A, Lingwood CA. Induction of verotoxin sensitivity in receptor-deficient cell lines using the receptor glycolipid globotriosylceramide. Proc Natl Acad Sci USA 1990; 87:7898-7901.

47. Maloney MD, Lingwood CA. CD19 has a potential CD77 (globotriaosyl ceramide)-binding site with sequence similarity to verotoxin B-subunits: Implications of molecular mimicry for B cell adhesion and enterohemorrhagic Escherichia coli pathogenesis. J Exp Med 1994; 180:191-201.

48. Madassery JV, Gillard B, Marcus DM, Nahm MH. Subpopulations of B cells in germinal centers. III. HJ6, a monoclonal antibody, binds globoside and a subpopulation of germinal center B cells. J Immunol 1991; 147:823-829.

49. Wiels J, Mangeney M, Tetaud C, Tursz T. Sequential shifts in the three major glycosphingolipid series are associated with B cell differentiation. International Immunology 1991; 3:1289-1300.

50. Taga S, Tetaud C, Mangeney M, Tursz T, Wiels J. Sequential changes in glycolipid expression during human B cell differentiation: Enzymatic bases. Biochimica et Biophysica Acta 1995; 1254:56-65.

51. Kansas GS, Wood GS, Tedder TF. Expression, distribution, and biochemistry of human CD39. Role in activation-associated homotypic adhesion of lymphocytes. J Immunol 1991; 146:2235-2244.

52. Pascual V, Liu YJ, Magalski A, de Bouteiller O, Banchereau J, Capra et al. Analysis of somatic mutation in five B cell subsets of human tonsil. J Exp Med 1994; 180:329-339.

53. Liu YJ, Barthelemy C, de Bouteiller O, Arpin C, Durand I, Banchereau et al. Memory B cells from human tonsils colonize mucosal epithelium and directly present antigen to T cells by rapid up-regulation of B7-1 and B7-2. Immunity 1995; 2:239-248.

54. Nahm MH, Takes PA, Bowen MB, Macke KA. Subpopulations of B lymphocytes in germinal centers, II. A germinal center B cell subpopulation expresses sIgD and CD23. Immunology Letters 1989; 21:201-208.

55. Billian G, Bella C, Mondiere P, Defrance T. Identification of a tonsil IgD$^+$ B cell subset with phenotypical and functional characteristics of germinal center B cells. Eur J Immunol 1996; 26:1712-1719.

56. Liu YJ, de Bouteiller O, Arpin C, Briere F, Galibert L, Ho S et al. Normal human IgD$^+$IgM$^-$ germinal center B cells can express up to 80 mutations in the variable region of their IgD transcripts. Immunity 1996; 4:603-613.

57. Papavasiliou F, Casellas R, Suh HY, Qin XF, Besmer E, Pelanda R et al. V(D)J recombination in mature B cells—a mechanism for altering antibody responses. Science 1997; 278:298-301.

58. Han SH, Dillon SR, Zheng B, Shimoda M, Schlissel MS, Kelsoe G. V(D)J recombinase activity in a subset of germinal center B lymphocytes. Science 1997; 278:301-305.

59. Han S, Zheng B, Schatz DG, Spanopoulou E, Kelsoe G. Neoteny in lymphocytes: Rag1 and Rag2 expression in germinal center B cells. Science 1996; 274:2094-2097.

60. Hikida M, Mori M, Takai T, Tomochika K, Hamatani K, Ohmori H. Reexpression of RAG-1 and RAG-2 genes in activated mature mouse B cells. Science 1996; 274:2092-2094.

61. Liu YJ, Malisan F, de Bouteiller O, Guret C, Lebecque S, Banchereau et al. Within germinal centers, isotype switching of immunoglobulin genes occurs after the onset of somatic mutation. Immunity 1996; 4:241-250.

62. Lederman S, Yellin MJ, Inghirami G, Lee JJ, Knowles DM, Chess L. Molecular interactions mediating T-B lymphocyte collaboration in human lymphoid follicles. Roles of T cell-B-cell-activating molecule (5c8 antigen) and CD40 in contact-dependent help. J Immunol 1992; 149:3817-3826.

63. Butch AW, Chung GH, Hoffmann JW, Nahm MH. Cytokine expression by germinal center cells. J Immunol 1993; 150:39-47.

64. Feuillard J, Taylor D, Casamayor-Palleja M, Johnson GD, MacLennan IC. Isolation and characteristics of tonsil centroblasts with reference to Ig class switching. International Immunology 1995; 7:121-130.

65. Weiser P, Muller R, Braun U, Reth M. Endosomal targeting by the cytoplasmic tail of membrane immunoglobulin. Science 1997; 276:407-409.

66. Achatz G, Nitschke L, Lamers MC. Effect of transmembrane and cytoplasmic domains of IgE on the IgE response. Science 1997; 276:409-411.

67. Kaisho T, Schwenk F, Rajewsky K. The roles of gamma-1 heavy chain membrane expression and cytoplasmic tail in IgG1 responses. Science 1997; 276:412-415.

68. Hathcock KS, Hodes RJ. Role of the CD28-B7 costimulatory pathways in T cell-dependent B cell responses. [Review] [152 refs]. Adv Immunol 1996; 62:131-166.

69. Vyth-Dreese FA, Dellemijn TA, Majoor D, de JD. Localization in situ of the co-stimulatory molecules B7.1, B7.2, CD40 and their ligands in normal human lymphoid tissue. Eur J Immunol 1995; 25:3023-3029.

70. Tarlinton D. Germinal centers-a second childhood for lymphocytes. Current Biology 1997; 7:R 155-R 159

71. Hikida M, Mori M, Kawabata T, Takai T, Ohmori H. Characterization of B cells expressing recombination activating genes in germinal centers of immunized mouse lymph nodes. J Immunol 1997; 158:2509-2512.

72. Radic MZ, Zouali M. Receptor editing, immune diversification, and self-tolerance. [Review] [50 refs]. Immunity 1996; 5:505-511.

73. Hu BT, Lee SC, Marin E, Ryan DH, Insel RA. Telomerase is up-regulated in human germinal center B cells in vivo and can be re-expressed in memory B cells activated in vitro. J Immunol 1997; 159:1068-1071.

74. Weng NP, Granger L, Hodes RJ. Telomere lengthening and telomerase activation during human B cell differentiation. Proc Natl Acad Sci USA 1997; 94:10827-10832.

75. deSaint-Vis B, Cupillard L, Pandrau-Garcia D, Ho S, Renard N, Grouard G et al. Distribution of carboxypeptidase M on lymphoid and myeloid cells parallels the other zinc-dependent proteases CD10 and CD13. Blood 1995; 86:1098-1105.

76. Gregory CD, Tursz T, Edwards CF, Tetaud C, Talbot M, Caillou B et al. Identification of a subset of normal B cells with a Burkitt's lymphoma (BL)-like phenotype. J Immunol 1987; 139:313-318.

77. Kalled SL, Siva N, Stein H, Reinherz EL. The distribution of CD10 (NEP 24.11, CALLA) in humans and mice is similar in non-lymphoid organs but differs within the hematopoietic system: Absence on murine T and B lymphoid progenitors. Eur J Immunol 1995; 25:677-687.

78. Woronicz JD, Calnan B, Ngo V, Winoto A. Requirement for the orphan steroid receptor Nur77 in apoptosis of T-cell hybridomas. Nature 1994; 367:277-281.

79. Woronicz JD, Lina A, Calnan BJ, Szychowski S, Cheng L, Winoto A. Regulation of the Nur77 orphan steroid receptor in activation-induced apoptosis. Mol Cell Biol 1995; 15:6364-6376.

80. Koopman G, Keehnen RM, Lindhout E, Newman W, Shimizu Y, van Seventer et al. Adhesion through the LFA-1 (CD11a/CD18)-ICAM-1 (CD54) and the VLA-4 (CD49d)-VCAM-1 (CD106) pathways prevents apoptosis of germinal center B cells. J Immunol 1994; 152:3760-3767.

81. Liu YJ, Grouard G, de Bouteiller O, Banchereau J. Follicular dendritic cells and germinal centers. [Review] [205 refs]. International Review of Cytology 1996; 166:139-179.

82. Liu YJ, Joshua DE, Williams GT, Smith CA, Gordon J, MacLennan IC. Mechanism of antigen-driven selection in germinal centres. Nature 1989; 342:929-931.

83. Kremmidiotis G, Zola H. Changes in CD44 expression during B cell differentiation in the human tonsil. Cell Immunol 1995; 161:147-157.

84. Hathcock KS, Hirano H, Murakami S, Hodes RJ. CD44 expression on activated B cells. Differential capacity for CD44-dependent binding to hyaluronic acid. J Immunol 1993; 151:6712-6722.

85. Koopman G, Griffioen AW, Ponta H, Herrlich P, van dB, Manten-Horst E et al. CD44 splice variants; expression on lymphocytes and in neoplasia. [Review] [24 refs]. Res Immunol 1993; 144:750-754.

86. Liu YJ, Cairns JA, Holder MJ, Abbot SD, Jansen KU, Bonnefoy JY et al. Recombinant 25-kDa CD23 and interleukin 1 alpha promote the survival of germinal center B cells: Evidence for bifurcation in the development of centrocytes rescued from apoptosis. Eur J Immunol 1991; 21:1107-1114.

87. Bonnefoy JY, Henchoz S, Hardie D, Holder MJ, Gordon J. A subset of anti-CD21 antibodies promote the rescue of germinal center B cells from apoptosis. Eur J Immunol 1993; 23:969-972.

88. van der Voort R, Taher TE, Keehnen RM, Smit L, Groenink M, Pals ST. Paracrine regulation of germinal center B cell adhesion through the c-Met-hepatocyte growth factor/scatter factor pathway. J Exp Med 1997; 185:2121-2131.

89. Krenacs T, van Dartel M, Lindhout E, Rosendaal M. Direct cell/cell communication in the lymphoid germinal center: Connexin43 gap junctions functionally couple follicular dendritic cells to each other and to B lymphocytes. Eur J Immunol 1997; 27:1489-1497.

90. Airas L, Jalkanen S. CD73 mediates adhesion of B cells to follicular dendritic cells. Blood 1996; 88:1755-1764.

91. Reichert RA, Gallatin WM, Weissman IL, Butcher EC. Germinal center B cells lack homing receptors necessary for normal lymphocyte recirculation. J Exp Med 1983; 157:813-827.

92. Martinez-Valdez H, Guret C, de Bouteiller O, Fugier I, Banchereau J, Liu YJ. Human germinal center B cells express the apoptosis-inducing genes Fas, c-Myc, P53, and Bax but not the survival gene bcl-2. J Exp Med 1996; 183:971-977.

93. Cutrona G, Dono M, Pastorino S, Ulivi M, Burgio VL, Zupo S et al. The propensity to apoptosis of centrocytes and centroblasts correlates with elevated levels of intracellular myc protein. Eur J Immunol 1997; 27:234-238.

94. Cutrona G, Ulivi M, Fais F, Roncella S, Ferrarini M. Transfection of the c-Myc oncogene into normal Epstein-Barr virus-harboring B cells results in new phenotypic and functional features resembling those of Burkitt lymphoma cells and normal centroblasts. J Exp Med 1995; 181:699-711.

95. Gray D, Siepmann K, van Essen D, Poudrier J, Wykes M, Jainandunsing et al. B-T lymphocyte interactions in the generation and survival of memory cells. [Review] [65 refs]. Immunol Rev 1996; 150:45-61.

96. Foy TM, Aruffo A, Bajorath J, Buhlmann JE, Noelle RJ. Immune regulation by CD40 and its ligand GP39. [Review] [111 refs]. Ann Rev Immunol 1996; 14:591-617.

97. Foy TM, Laman JD, Ledbetter JA, Aruffo A, Claassen E, Noelle RJ. gp39-CD40 interactions are essential for germinal center formation and the development of B cell memory. J Exp Med 1994; 180:157-163.

98. Lindhout E, Lakeman A, de GC. Follicular dendritic cells inhibit apoptosis in human B lymphocytes by a rapid and irreversible blockade of preexisting endonuclease. J Exp Med 1995; 181:1985-1995.

99. Koopman G, Keehnen RM, Lindhout E, Zhou DF, de Groot C, Pals ST. Germinal center B cells rescued from apoptosis by CD40 ligation or attachment to follicular dendritic cells, but not by engagement of surface immunoglobulin or adhesion receptors, become resistant to CD95-induced apoptosis. Eur J Immunol 1997; 27:1-7.

100. Cleary AM, Fortune SM, Yellin MJ, Chess L, Lederman S. Opposing roles of CD95 (Fas/APO-1) and CD40 in the death and rescue of human low density tonsillar B cells. J Immunol 1995; 155:3329-3337.

101. Watanabe D, Suda T, Nagata S. Expression of Fas in B cells of the mouse germinal center and Fas-dependent killing of activated B cells. International Immunology 1995; 7:1949-1956.

102. Lagresle C, Mondiere P, Bella C, Krammer PH, Defrance T. Concurrent engagement of CD40 and the antigen receptor protects naive and memory human B cells from APO-1/Fas-mediated apoptosis. J Exp Med 1996; 183:1377-1388.

103. Gordon J, Gregory CD, Grafton G, Pound JD. Signals for survival and apoptosis in normal and neoplastic B lymphocytes. [Review] [10 refs]. Adv Exp Med & Biol 1996; 406:139-144.

104. Schattner EJ, Elkon KB, Yoo DH, Tumang J, Krammer PH, Crow MK et al. CD40 ligation induces Apo-1/Fas expression on human B lymphocytes and facilitates apoptosis through the Apo-1/Fas pathway. J Exp Med 1995; 182:1557-1565.

105. Garrone P, Neidhardt EM, Garcia E, Galibert L, Van Kooten C, Banchereau J. Fas ligation induces apoptosis of CD40-activated human B lymphocytes. J Exp Med 1995; 182:1265-1273.

106. Billian G, Mondiere P, Berard M, Bella C, Defrance T. Antigen receptor-induced apoptosis of human germinal center B cells is targeted to a centrocytic subset. Eur J Immunol 1997; 27:405-414.

107. Galibert L, Burdin N, Barthelemy C, Meffre G, Durand I, Garcia E et al. Negative selection of human germinal center B cells by prolonged BCR crosslinking. J Exp Med 1996; 183:2075-2085.

108. Hahne M, Renno T, Schroeter M, Irmler M, French L, Bornard T et al. Activated B cells express functional Fas ligand. Eur J Immunol 1996; 26:721-724.

109. Dunn-Walters DK, Isaacson PG, Spencer J. Analysis of mutations in immunoglobulin heavy chain variable region genes of microdissected marginal zone (MGZ) B cells suggests that the MGZ of human spleen is a reservoir of memory B cells. J Exp Med 1995; 182:559-566.

110. Dono M, Burgio VL, Tacchetti C, Favre A, Augliera A, Zupo S et al. Subepithelial B cells in the human palatine tonsil. I. Morphologic, cytochemical and phenotypic characterization. Eur J Immunol 1996; 26:2035-2042.

111. Dono M, Zupo S, Augliera A, Burgio VL, Massara R, Melagrana A et al. Subepithelial B cells in the human palatine tonsil. II. Functional characterization. Eur J Immunol 1996; 26:2043-2049.

112. Gray D. Immunological memory. [Review] [201 refs]. Annual Review of Immunology 1993; 11:49-77.
113. Ahmed R, Gray D. Immunological memory and protective immunity: Understanding their relation. [Review] [75 refs]. Science 1996; 272:54-60.
114. Bachmann MF, Odermatt B, Hengartner H, Zinkernagel RM. Induction of long-lived germinal centers associated with persisting antigen after viral infection. J Exp Med 1996; 183:2259-2269.
115. Arpin C, Dechanet J, Van Kooten C, Merville P, Grouard G, Briere F et al. Generation of memory B cells and plasma cells in vitro. Science 1995; 268:720-722.
116. Arpin C, Banchereau J, Liu YJ. Memory B cells are biased towards terminal differentiation—a strategy that may prevent repertoire freezing. J Exp Med 1997; 186:931-940.
117. Neurath MF, Stuber ER, Strober W. BSAP: A key regulator of B-cell development and differentiation. [Review] [22 refs]. Immunol Today 1995; 16:564-569.
118. Wakatsuki Y, Neurath MF, Max EE, Strober W. The B cell-specific transcription factor BSAP regulates B cell proliferation. J Exp Med 1994; 179:1099-1108.
119. Max EE, Wakatsuki Y, Neurath MF, Strober W. The role of BSAP in immunoglobulin isotype switching and B-cell proliferation. Current Topics in Microbiology & Immunology 1995; 194:449-458.
120. Usui T, Wakatsuki Y, Matsunaga Y, Kaneko S, Kosek H, Kita T. Overexpression of B cell-specific activator protein (BSAP/PAX-5) in a late B cell is sufficient to suppress differentiation to an Ig high producer cell with plasma cell phenotype. J Immunol 1997; 158:3197-3204.
121. Rinkenberger JL, Wallin JJ, Johnson KW, Koshland ME. An interleukin-2 signal relieves BSAP (Pax-5)-mediated repression of the immunoglobulin J chain gene. Immunity 1996; 5:377-386.
122. Michaelson JS, Singh M, Birshtein BK. B cell lineage-specific activator protein (BSAP). A player at multiple stages of B cell development. [Review] [50 refs]. J Immunol 1996; 156:2349-2351.
123. Zwollo P, Arrieta H, Ede K, Molinder K, Desiderio S, Pollock R. The Pax-5 gene is alternatively spliced during B-cell development. J Biol Chem 1997; 272:10160-10168.
124. Turner CAJ, Mack DH, Davis MM. Blimp-1, a novel zinc finger-containing protein that can drive the maturation of B lymphocytes into immunoglobulin-secreting cells. Cell 1994; 77:297-306.
125. Huang S. Blimp-1 is the murine homolog of the human transcriptional repressor PRDI-BF1 [letter]. Cell 1994; 78:9
126. Schliephake DE, Schimpl A. Blimp-1 overcomes the block in IgM secretion in lipopolysaccharide/anti-mu F(ab')2-co-stimulated B lymphocytes. Eur J Immunol 1996; 26:268-271.
127. Lin Y, Wong K, Calame K. Repression of c-Myc transcription by Blimp-1, an inducer of terminal B cell differentiation. Science 1997; 276:596-599.

128. Onizuka T, Moriyama M, Yamochi T, Kuroda T, Kazama A, Kanazawa N et al. Bcl-6 gene product, a 92- to 98-kD nuclear phosphoprotein, is highly expressed in germinal center B cells and their neoplastic counterparts. Blood 1995; 86:28-37.

129. Cattoretti G, Chang CC, Cechova K, Zhang J, Ye BH, Falini B et al. Bcl-6 protein is expressed in germinal-center B cells. Blood 1995; 86:45-53.

130. Allman D, Jain A, Dent A, Maile RR, Selvaggi T, Kehry MR et al. Bcl-6 expression during B-cell activation. Blood 1996; 87:5257-5268.

131. Dent AL, Shaffer AL, Yu X, Allman D, Staudt LM. Control of inflammation, cytokine expression, and germinal center formation by Bcl-6. Science 1997; 276:589-592.

132. Fukuda T, Yoshida T, Okada S, Hatano M, Miki T, Ishibashi K et al. Disruption of the BCL6 gene results in an impaired germinal center formation. J Exp Med 1997; 186:439-448.

133. Katz P, Whalen G, Kehrl JH. Differential expression of a novel protein kinase in human B lymphocytes. Preferential localization in the germinal center. J Biol Chem 1994; 269:16802-16809.

134. Christoph T, Rickert R, Rajewsky K. M17: A novel gene expressed in germinal centers. International Immunology 1994; 6:1203-1211.

135. Weinberg DS, Ault KA, Gurley M, Pinkus GS. The human lymph node germinal center cell: Characterization and isolation by using two-color flow cytometry. J Immunol 1986; 137:1486-1494.

136. Mathis DJ, Benoist CO, Williams VE, Kanter MR, McDevitt HO. The murine E alpha immune response gene. Cell 1983; 32:745-754.

137. van EW, Ron Y, Monaco J, Kappler J, Marrack P, Le MM et al. Compartmentalization of MHC class II gene expression in transgenic mice. Cell 1988; 53:357-370.

138. Kuo FC, Sklar J. Augmented expression of a human gene for 8-oxoguanine DNA glycosylase (MutM) in B lymphocytes of the dark zone in lymph node germinal centers. J Exp Med 1997; 186:1547-1556.

139. Delibrias CC, Floettmann JE, Rowe M, Fearon DT. Downregulated expression of SHP-1 in Burkitt lymphomas and germinal center B lymphocytes. J Exp Med 1997; 186:1575-1583.

140. To SS, Magoulas T, Nicholson E, Schrieber L. Identification of a human endothelial cell activation antigen that is co-expressed by germinal follicle centre B lymphocytes. Immunol 1992; 76:616-624.

141. Epstein AL, Marder RJ, Winter JN, Fox RI. Two new monoclonal antibodies (LN-1, LN-2) reactive in B5 formalin-fixed, paraffin-embedded tissues with follicular center and mantle zone human B lymphocytes and derived tumors. J Immunol 1984; 133:1028-1036.

142. Powell LD, Sgroi D, Sjoberg ER, Stamenkovic I, Varki A. Natural ligands of the B cell adhesion molecule CD22 beta carry N-linked oligosaccharides with alpha-2,6-linked sialic acids that are required for recognition. J Biol Chem 1993; 268:7019-7027.

143. Jacobson EB, Caporale LH, Thorbecke GJ. Effect of thymus cell injections on germinal center formation in lymphoid tissues of nude (thymusless) mice. Cell Immunol 1974; 13:416-430.

144. Weissman IL, Gutman GA, Friedberg SH, Jerabek L. Lymphoid tissue architecture. III. Germinal centers, T cells, and thymus-dependent vs thymus-independent antigens. Adv Exp Med and Biol 1976; 66:229-237.
145. Vonderheide RH, Hunt SV. Does the availability of either B cells or CD4$^+$ cells limit germinal centre formation? Immunol 1990; 69:487-489.
146. Wang D, Wells SM, Stall AM, Kabat EA. Reaction of germinal centers in the T-cell-independent response to the bacterial polysaccharide alpha(1→6)dextran. Proc Natl Acad Sci USA 1994; 91:2502-2506.
147. Davies AJ, Carter RL, Leuchars E, Wallis V, Dietrich FM. The morphology of immune reactions in normal, thymectomized and reconstituted mice III. Response to bacterial antigens: Salmonellar flagellar antigen and pneumococcal polysaccharide. Immunol 1970; 19:945-957.
148. Gutman GA, Weissman IL. Lymphoid tissue architecture experimental analysis of the origin and distributation of T-cells and B-cells. Immunol 1972; 23:465-479.
149. Si L, Roscoe G, Whiteside TL. Selective distribution and quantitation of T-lymphocyte subsets in germinal centers of human tonsils. Arch Pathol Lab Med 1983; 107:228-231.
150. Brachtel EF, Washiyama M, Johnson GD, Tenner-Racz K, Racz P, MacLennan IC. Differences in the germinal centres of palatine tonsils and lymph nodes. Scandinavian Journal of Immunology 1996; 43:239-247.
151. Bowen MB, Butch AW, Parvin CA, Levine A, Nahm MH. Germinal center T cells are distinct helper-inducer T cells. Hum Immunol 1991; 31:67-75.
152. Rouse RV, Ledbetter JA, Weissman IL. Mouse lymph node germinal centers contain a selected subset of T cells-the helper phenotype. J Immunol 1982; 128:2243-2246.
153. Poppema S, Bhan AK, Reinherz EL, McCluskey RT, Schlossman SF. Distribution of T cell subsets in human lymph nodes. J Exp Med 1981; 153:30-41.
154. Cosgrove D, Gray D, Dierich A, Kaufman J, Lemeur M, Benoist C et al. Mice Lacking MHC Class II Molecules. Cell 1991; 66:1051-1066.
155. Pulido R, Cebrian M, Acevedo A, de Landazuri MO, Sanchez-Madrid F. Comparative biochemical and tissue distribution study of four distinct CD45 antigen specificities. J Immunol 1988; 140:3851-3857.
156. Dianda L, Gulbranson-Judge A, Pao W, MacLennan IC, Owen MJ. Germinal center formation in mice lacking αβ T cells. Eur J Immunol 1996; 26:1603-1607.
157. Lederman S, Yellin MJ, Inghirami G, Lee JJ, Knowles DM, Chess L. Molecular interactions mediating T-B lymphocyte collaboration in human lymphoid follicles. Roles of T cell-B-activating molecule (5c8 Antigen) and CD40 in contact-dependent help. J Immunol 1992; 149:3817-3826.

158. van den Eertwegh AJ, Noelle RJ, Roy M, Shepherd DM, Aruffo A, Ledbetter JA. In vivo CD40-gp39 interactions are essential for thymus-dependent humoral immunity. I. In vivo expression of CD40 ligand, cytokines, and antibody production delineates sites of cognate T-B cell interactions. J Exp Med 1993; 178:1555-1565.

159. Ferguson SE, Han S, Kelsoe G, Thompson CB. CD28 is required for germinal center formation. J Immunol 1996p 156:4576-4581.

160. Lane P, Burdet C, Hubele S, Scheidegger D, Muller U, McConnell F et al. B cell function of mice transgenic for mCTLA-H gamma 1: Lack of germinal centers correlated with poor affinity maturation and class switching despite normal priming of CD4$^+$ T cells. J Exp Med 1994; 179:819-830.

161. Wen L, Pao W, Wong FS, Peng Q, Craft J, Zheng B et al. Germinal center formation, immunoglobulin class switching, and autoantibody production driven by "non alpha/beta" T cells. J Exp Med 1996; 183:2271-2282.

162. Bruno L, Rocha B, Rolink A, von Boehmer H, Ridewabdk HR. Intra- and extra-thymic expression of the pre-T cell receptor alpha gene. Eur J Immunol 1995; 25:1877-1882.

163. Philpott KL, Viney JL, Kay G, Rastan S, Gardiner EM, Chae S et al. Lymphoid development in mice congenitally lacking T cell receptor αβ-expressing cells. Science 1992; 256:1448-1452.

164. Fuller KA, Kanagawa O, Nahm MH. T cells within germinal centers are specific for the immunizing antigen. J Immunol 1993; 151:4505-4512.

164a. Zheng B, Han S, Kelsoe G. T helper cells in murine germinal centers are antigen-specific emigrants that downregulate Thy-1. J Exp Med 1996; 184:1083-1091.

165. Zheng B, Han S, Zhu Q, Goldsby R, Kelsoe G. Alternative pathways for the selection of antigen-specific peripheral T cells. Nature 1996; 384:263-266.

166. Gulbranson-Judge A, MacLennan I. Sequential antigen-specific growth of T cells in the T zones and follicles in response to pigeon cytochrome *c*. Eur J Immunol 1996; 26:1830-1837.

167. Berman MA, Rafiei S, Gutman GA. Association of T cells with proliferating cells in lymphoid follicles. Transplantation 1981; 32:426-430.

168. Kelly KA, Bucy RP, Nahm MH. Germinal center T cells exhibit properties of memory helper T cells. Cell Immunol 1995; 163:206-214.

169. Pape KA, Kearney ER, Khoruts A, Mondino A, Merica R, Chen ZM et al. Use of adoptive transfer of T-cell-antigen-receptor-transgenic T cells for the study of T cell activation in vivo. Immunol Rev 1997; 156:67-78.

170. Kearney ER, Pape KA, Loh DY, Jenkins MK. Visualization of peptide-specific T cell immunity and peripheral tolerance induction in vivo. Immunity 1994; 1:327-339.

171. Zheng B, Xue W, Kelsoe G. Locus-specific somatic hypermutation in germinal centre T cells. Nature 1994; 372:556-559.

172. Poppema S, Visser L, De Leij L. Reactivity of presumed anti-natural killer cell antibody Leu 7 with intrafollicular T lymphocytes. Clin Exp Immunol 1983; 54:834-837.

173. Chou DK, Ilyas AA, Evans JE, Costello C, Quarles RH, Jungalwala FB. Structure of sulfated glucuronyl glycolipids in the nervous system reacting with HNK-1 antibody and some IgM paraproteins in neuropathy. J Biol Chem 1986; 261:11717-11725.

174. Needham LK, Schnaar RL. The HNK-1 reactive sulfoglucuronyl glycolipids are ligands for L-selectin and P-selectin but not E-selectin. Proc Natl Acad Sci USA 1993; 90:1359-1363.

175. McHeyzer-Williams MG, Davis MM. Antigen-specific development of primary and memory T cells in vivo. Science 1995; 268:106-111.

176. Nieuwenhuis P, Opstelten D. Functional anatomy of germinal centers. Am J Anat 1984; 170:421-435.

177. Grouard G, Durand I, Filgueira L, Banchereau J, Liu YJ. Dendritic cells capable of stimulating T cells in germinal centres. Nature 1996; 384:364-367.

Somatic Hypermutation of Immunoglobulin Genes

U. Storb

Brief Historic Overview

The first convincing evidence for somatic mutation of Ig genes was a study of mouse λ genes carried out by Weigert and collaborators in the early 70s.[1] Comparing the amino acid sequences of λ light chain variable (V) regions in nine mouse myelomas, they found that six were identical, and three had one, two, or three amino acid changes. They concluded that there was only a single variable gene for mouse lambda, and that the three altered sequences must have arisen by some process of somatic mutation.

In the late 70s Tonegawa's laboratory showed that functional light chain genes are created by recombination on the DNA level.[2] By determining the sequence of a mouse λ variable (V) gene in embryonic DNA, as a source of germline genes, and comparing it with the DNA sequence in one of the λ producing myelomas that Weigert and colleagues had studied, they confirmed the conclusions of the earlier study. Somatic mutation was further verified on the DNA level by a more extensive study of light and heavy chain V genes from several mouse myelomas.[3,4] It was noted that the mutations are present not only in the hypervariable regions, but also in the framework, as well as in the introns of Ig genes, suggesting that the mechanism may not recognize functional domains of the Ig protein. Furthermore, germline, unrearranged V genes were found to be unmutated, which implied that somatic mutation did not occur before V(D)J recombination.[5]

The Biology of Germinal Centers in Lymphoid Tissue, edited by G. Jeanette Thorbecke and Vincent K. Tsiagbe. © 1998 Springer-Verlag and R.G. Landes Company.

Biology of Somatic Hypermutation

While V(D)J recombination occurs during development of the preB cell stages in the bone marrow, there is no experimental evidence that the somatic mutation process is normally active during early B cell development. Thus, major changes from germline V, (D), and J genes in heavy and light chain genes expressed in preB cells are located at VD and DJ junctions.[6,7] They can be explained as arising from insertions of untemplated nucleotides (N regions) by terminal transferase, or of palindromic nucleotides (P) during hairpin formation, or from deletions during the processing of the coding ends.[8] However, untransformed preB cells have not been systematically studied for somatic mutation. A preB cell line has been described that in culture undergoes a low frequency of point mutations in the variable (V) and constant (C) regions.[9,10] The mutation frequency is increased after fusion with a myeloma cell line.[9] Since this preB cell line, 18.81, also undergoes switch recombination,[11] a property normally seen only in activated, mature B cells, it may have activated gene expression programs that are normally repressed in preB cells.

Mature B cells in the periphery that co-express IgM and IgD generally have only rearranged germline V genes.[12,13] It is generally assumed that Ig synthesis is downregulated as B cells initiate germinal center formation;[14] however there is some controversy as to whether or not cells in the dark zone produce Ig.[15] In rat, IgD⁻ B cells form germinal centers much more efficiently than IgD⁺ cells.[16] Curiously, a population of centroblasts in the dark zone of human tonsils has been found to express only IgD and to have suffered extremely high levels of somatic mutation.[17] Their role is unclear, but apparently there is a high frequency of somatic mutation in human IgM⁻, IgD⁺ plasma cells and myelomas which may be the progeny of IgD⁺ germinal center cells (S. Lebeque, personal communication).

Somatic mutation appears to be induced after mature B cells have been activated by antigen and have interacted with T-helper cells to initiate a germinal center. The B cells appear to originate from periarteriolar lymphocyte sheaths (PALs) in which other B cells, similarly activated by antigen and T cell help, can undergo class switch recombination. It is not clear how the events of somatic mutation and switching are controlled in individual B cells, since both occur after antigen stimulation and interaction with T-helper cells. Apparently somatic mutation only occurs in the environment of the germinal center, but class switching can occur in germinal centers as well as in the follicles.[18] Class switch recombination does not terminate somatic mutation.[19] Thus, B cells can somatically mutate before as well as after switching and one finds IgM genes that are mutated and IgG genes that are unmutated. In general, however, a greater proportion of cells producing IgG than IgM have mutations in their heavy and/or light chain genes, and the number of mutations per gene is higher in IgG[+] cells.[20-23] It is not clear whether the mutation process is more active in IgG producing cells, or if these cells remain longer in the mutating compartment, or reenter it one or more times (see below).

The mutation process appears to occur in the dark zone of the germinal center where B cells proliferate at a high rate with an average cycling time of 6 to 7 hours,[14] one of the highest in adult vertebrate cells. After presumably only a few cell generations, the B cells migrate into the light zone of the germinal center where selection of cells carrying functional Ig occurs,[24,25] presumably by interaction with follicular dendritic cells presenting the immunizing antigen in undegraded form on their cell surface.[26] B cells that have undergone somatic mutation leading to deficient heavy and/or light chain genes are believed to die by apoptosis.[27] Thus, B cells with changes in Igs that lead to a low affinity for the antigen or to the loss of membrane Ig altogether, due to an in frame stop codon or changes in the Ig framework that prevent proper alignment of the beta pleated sheets, are counterselected. B cells with mutations that are not highly selectable and that are normally not seen in the long lived B cell pool have been found among single cells picked from germinal centers,[28] supporting the idea that such cells are not immediately selected against.

Whether B cells with mutations that have been positively selected in the light zone reenter the dark zone one or multiple times, or whether somatic mutation can also occur outside of the dark zone is an open question. Recent data by C. Berek suggest that somatic mutation is not restricted to the dark zone and that the overall pattern of development of B cells is more complex than has been described so far (C. Berek, personal communication). Clearly, memory B cells with mutations are created at some time in the process. This may take place continuously during the life of a germinal center (up to about three to four weeks after immunization, depending on the immunization scheme[27]), or memory cells may exit into the circulating pool of B cells only as the germinal center wanes.

A single germinal center appears to represent the progeny of only one to three precursor B cells. This has been deduced from immunizations with two different antigens, where individual germinal centers most often reacted with only one of the antigens.[29] Furthermore, B cells within one germinal center have been found to be clonally related when single cells were picked from a single germinal center:[28] Multiple cells carried the same VDJ joints, including N regions, in their heavy chain genes. Their VH genes could therefore be related by constructing a genealogical tree of base substitutions. Interestingly, only the terminal branches of such trees were recovered, but not most of the intermediate steps leading to these terminal somatic mutations. This suggests that the process of somatic mutation is continuous and rapid, so that the time of retaining a newly created mutation before a new one is added is short. In addition, there may be a continuous selection of cells with the elimination of those that carry V regions that have been mutated in certain unfavorable ways. However, if such continuous selection exists, it cannot be radical, since none of the single cells that were isolated from germinal centers after immunization with (4-hydroxy-3-nitrophenyl) acetyl (NP) carried mutations characteristic of affinity maturation, a tryptophane to leucine exchange in amino acid position 33 of the heavy chain.[28] This mutation is present in 70% of hybridomas from NP hyperimmunized mice. On the other hand, there is additional support for continuous selection: Cells that were isolated from germinal centers showed a low ratio of replacement/silent (R/S) mutations in the framework encoding regions (FR) of the heavy chain genes.[28] This is to be expected if the cells which produce Ig molecules that cannot fold into proper Ig conformation are rapidly eliminated. A different interpretation for the lack of devastating changes in the framework of mutated Ig molecules was given by Reynaud/Weill and colleagues in studies with sheep Igs.[30] Fetal and newborn lambs show a high level of somatic hypermutation in the ileal Peyer's patches. The distribution of the mutations in the Ig genes shows a high R/S ratio in the complementarity determining regions (CDRs), but not the framework. Reynaud et al concluded that this must be a result of the mutation mechanism per se, since the same high R/S ratio was found in Ig genes of newborn lambs who had ileal loops that had been surgically isolated at the early fetal stage so that they had no contact with intestinal antigens, and in germfree animals. This was interpreted as indicating that the primary sequences of the CDRs had been selected over evolutionary time to represent sequences that are highly mutable and therefore susceptible to change. It was further postulated that the framework sequences had evolved to be resistant to mutation and thus tended to be conserved. However, it is possible that even without antigen selection mutations in the framework are selected against. It is likely that B cells require Ig receptors for survival. This was dramatically demonstrated by Rajewsky's laboratory with mice whose Cμ gene was deleted by Cre/Lox targeting in adult B cells: No B cells were found that lacked Ig (personal communication).

The elimination of B cells that produce mutated Ig molecules that do not properly fold could occur without the expression of the Ig on the cell surface, for example by a chaperone signaling mechanism, or by poisoning the cells with an overload of defective protein. However, there seems to be no way to eliminate such cells without translating protein from the mutated Ig genes. While stop codons in internal exons of T cell receptor genes have been found to lead to nuclear instability of the mRNAs,[31] there is no evidence that this mechanism is linked to apoptosis. Thus one must assume that B cells in the germinal center translate the mutated Ig genes into protein at the time of selection against framework changes. Unless the assumption is correct that changes in the framework do not readily occur,[30] the selection must take place soon after creation of framework changes and thus most likely within the dark zone where mutations are believed to go on. However, there is conflicting evidence of whether Ig genes are expressed in the dark zone.[15] For the selection of high affinity antibody-producing cells, expression of the Ig on the cell membrane is required. Selection for high affinity may occur at a later time than framework selection, since cells that appeared to be selected for functional frameworks[28] did not have the high affinity trp to leu change (see above).

The process of somatic mutation must be downregulated before the germinal center involutes. B cells can reenter the cell cycle without further mutation as has been dramatically shown by Weigert and collaborators:[32] a large number of independent B cell hybridomas from the spleen of an autoimmune mouse showed many identical point mutations, but no different point mutations. Similar evidence of a secondary immune response involving B cells that may have mutated during the primary response, but did not acquire additional mutations upon reimmunization has been found by Rajewsky et al.[33] Both these observations made it questionable if memory B cells can reenter the somatic mutation process, or for that matter, can induce a germinal center reaction. On the other hand, with multiple immunizations, the proportion of Ig molecules with a high frequency of somatic mutation increases, suggesting that somatic mutation can be reactivated.[34,35] Alternatively, highly mutated sequences may become more and more selected.[36]

It was reported that somatic mutation occurs mainly in conventional B2 (CD5⁻) cells and that normal and malignant B1 (CD5⁺) cells show little or no somatic hypermutation (reviewed in ref. 37). However, others have found that changes due to somatic hypermutation are equally seen in human CD5⁺ and CD5⁻ B cells.[38] Also, human chronic lymphocytic leukemia cells are mainly CD5⁺ and have been found to contain heavy and light chain gene mutations.[39]

Besides B cells, T cells in the germinal center have been reported to have mutations in the TCR-α genes, but not TCR-β genes.[40] This report has so far not been confirmed, but in primary and memory T cells such mutations have not been observed.[41,42] Perhaps too few TCR genes have been sequenced in peripheral T cells, although, given the many complex interactions of the TCR, most random mutations in the TCR would be expected to be counterselected.

A number of mouse and human lymphoid cell lines contain mutated Ig genes. Mouse myelomas were the major source of cells that established that somatic mutations occur at a high frequency in the V regions of rearranged Ig genes.[2-4] Human multiple myelomas also show high levels of somatic mutation, but no intraclonal diversity, suggesting that each malignant clone had arisen from one mature B cell that had experienced residence in a germinal center.[43]

Both endemic and sporadic Burkitt's lymphoma cells show relatively highly mutated heavy and light chain genes, the former being mainly μ, rather than switched Ig classes.[44] Interestingly, there is a reduced R/S ratio in the framework region, suggesting that the cells had been selected for function. This finding implies that the malignant transformation occurred either in the germinal center after interaction with follicular dendritic cells or in memory B cells. Since the lymphomas are monoclonal, somatic mutation is not ongoing after transformation.

Monocytoid B cell lymphomas,[45] chronic lymphocytic leukemias[39] and follicular lymphomas[46] also show somatic mutation of Ig genes. The latter may continue somatic mutation in the malignant state,[46,47] which may serve the escape of the malignant cells from immune rejection.

Somatic Mutation Occurs in All Vertebrate Species

Somatic hypermutation has been observed in all vertebrates studied, although there exist a number of species-specific oddities. The highest frequencies have been found in human Igs, although, due to the high degree of polymorphism in the human population, it may be difficult sometimes to distinguish somatic mutation from unusual germline V genes. Somatic mutation is low or absent in IgM^+ /IgD^+ B cells in peripheral blood,[13] but present in IgM^+/IgD^- and IgM^-/IgD^- B cells. Evidence for somatic mutation was already observed in human fetal B cells[48] and was considerable in the blood B lymphocytes of a four year old child, with V region mutations in IgM of about 2%, and in IgG of about 3%.[21] In a 42 year old person, the mutation frequency in IgM was still about 2%, but the average frequency in Ig gamma genes was about 12% (Kim and Storb, unpublished), suggesting that IgG-producing memory B cells either reenter the mutation process, or continue to be selected for higher and higher affinity. IgM-producing cells on the other hand are apparently mainly derived from B cells that have recently developed. The best studied site of somatic mutation in the human is the tonsil, in which five B cell populations, Bm1 to Bm5, were distinguished. Based on surface markers, it was assumed that B cells progressed from Bm1 to Bm2 and so on to finally reach the Bm5 stage. Because very few or no somatic mutations were found in the Bm1 and Bm2 stages, it was assumed that somatic mutation is occurring in the transition from Bm2 (IgD^+, $CD23^+$) to Bm3 (IgD^-, $CD23^-$) stages.[22]

Mouse has the advantage of inbred strains, so that uncertainties due to polymorphism can be avoided. Most studies on the mechanism and biology of somatic mutation have been carried out in the mouse. Several model systems have been designed in the mouse (see next section). As in man, somatic mutation is a consequence of antigen stimulation.

Apparently without antigenic challenge, most of the somatic mutation processes in the sheep are ongoing during fetal and early postnatal life, up to about four months from birth, in the lymphoid tissues associated with the ileum and to a lesser degree the jejunum.[30,49,50] Since the spleen and lymph nodes also contain germinal centers, antigen driven mutation is presumably also going on in this species.

In the rabbit and the chicken, one major V(D)J rearrangement is used. The primordial light- and/or heavy-chain gene becomes highly diversified by a process of gene conversion that is ongoing early in life, and presumably not antigen dependent.[51,52] Gene conversion is most likely not a mechanism that is involved in somatic mutation in mouse and man (reviewed in ref. 53). On the other hand, Ig genes of rabbits and chicken, in addition to gene conversion, appear to incur somatic point mutations during stages of antigen induced diversification[54,55] (C. Thompson, personal communication).

Xenopus also shows strong evidence for somatic mutation.[56] However, there is little sign of affinity maturation and germinal centers are absent.[56] Thus, apparently, selection of the fittest B cells does not occur in this species.

A novel member of the Ig superfamily has been found in the nurse shark.[57] The nurse shark antigen receptor (NAR) molecule contains one variable and five constant domains and is found as a dimer in the serum. It resembles both Igs and TCRs and is highly diversified somatically in a way that suggests somatic hypermutation. NAR seems to be a novel Ig produced by B cells (M. Flajnik, personal communication).

Assay Systems to Study the Mechanism of Somatic Hypermutation

The early studies on somatic hypermutation were carried out by comparing the sequences of Ig genes expressed in myelomas of inbred Balb/c mice with the germline sequences of the same mouse strain (reviewed in ref. 58). Subsequently, mice were immunized with a variety of antigens, and somatic mutation was observed at varying times after primary or secondary antigen stimulation. In general, spleen B lymphocytes were immortalized as hybridomas for the isolation and sequencing of the rearranged Ig genes. Antigens that are well studied for induction of somatic mutation include the haptens phosphorylcholine (PC),[20,59] 2-phenyl-oxazolone (Ox),[34] 4-hydroxy-3-nitrophenylacetyl (NP),[60] and azophenylarsonate (Ars).[61]

As it became important to determine the cis-acting sequences that influence somatic mutation, the use of Ig transgenic mice was introduced.[62] Mice carrying a κ-light chain transgene derived from the myeloma MOPC167 with specificity for PC were hyperimmunized with PC and spleen hybridomas were isolated. In order to identify anti PC hybridomas derived from B cells that had undergone somatic mutation, the endogenous heavy chain mRNAs were sequenced. The κ-transgenes were then analyzed in hybridomas whose heavy chain genes were highly mutated. Somatic point mutations were found in the transgenic Vκ-regions of hybridomas from two different transgenic lines. The data clearly showed that cis-acting elements required to target the mutation process were present in the 16 kb long transgene and could function at different positions in the genome. As in somatic mutation of endogenous Ig genes, only the V region and its immediate flanks, but not the C region, were mutated, and the mutations showed a predominance of transitions over transversions.[62,63] Since multiple transgenes were present in these mice (three and twelve copies) and each one could, in principle, encode functional light chains, it could further be concluded that the targeting to the V region is an inherent property of the mutation process, and not due to selection against C region mutations.

The use of Ig transgenic mice as a source of B cell hybridomas (whose endogenous Ig genes could be sequenced to assess the cells' history of somatic mutation) was adopted as a convenient method for many studies of the mechanism of this mutation process.[23,64-74]

In order to avoid the possibility that cells were selected based on the state of expression of the transgene, transgenes with a stop codon within a leader or VJ exon were often used as passenger transgenes. The mRNAs transcribed from a transgene with a stop codon in a coding region were expected to be somewhat less stable, because they are not continuously protected by the translation machinery in the cytoplasm. Recent data have shown, however, that such stop codons already destabilize mRNAs in the nucleus.[75] It appears that this destabilization depends on the presence of an intron 3' of the exon containing the premature stop codon. Natural stop codons are located in the last exon. Thus, in order to be able to more accurately evaluate the mRNA levels transcribed from an Ig transgene, a stop codon would best be placed into the start of the last exon of interest, i.e., the C region exon in the case of kappa or lambda, or the last CH exon or the membrane exon in the case of secreted or membrane bound heavy chains, respectively.

Another breakthrough in the analysis of somatic mutation came with the perfection of a method to isolate germinal center B cells of mouse. It was based on the observation that germinal center B cells strongly bind the lectin peanut agglutinin (PNA).[76] Berek et al showed that, after immunization, the PNA[hi] B cells in germinal centers of the spleen accumulate somatic mutations.[25,77] Gonzalez-Fernandez and her collaborators further demonstrated that PNA[hi] /B220[+] B cells from intestinal Peyer's patches of unimmunized mice are highly enriched for cells which carry Ig gene somatic mutations, while PNA[lo]/B220[+] cells contain very few mutated sequences.[78,79] The mutations in Peyer's patches are assumed to be induced by chronic stimulation of B cells by antigens in the intestine, with the mutation frequency reaching a plateau around 5 months of age. The isolation of PNA^{hi}, B220$^+$ B cells from either the Peyer's patches of unimmunized mice or the spleen of hyperimmunized mice has become a very convenient standard method to enrich for B cells that have acquired somatic mutations in endogenous Ig genes or in transgenes.

In the study of somatic mutation in man, tonsils have been used as a tissue rich in germinal centers. Tonsillar B cells were separated into five fractions based on combinations of various surface markers.[22] Somatic mutation was studied in the Ig genes of these cell fractions and the conclusion was reached that mutation initiates and is ongoing in that fraction that shows clear evidence, but the lowest frequency, of somatic mutation. Peripheral blood lymphocytes were studied for somatic mutation either directly[21,80,81] or after immortalization by EBV.[82,83] Also, mRNAs and DNA from human spleen were found to contain Ig V sequences that were highly mutated.[84-86]

Most of the recent studies have made use of the PCR method using DNA or cDNA as a source of amplifiable genes. PCR also made it possible to analyze mutations in single cells isolated from germinal centers in microscopic sections of immunized mice.[28] When multiple cells were collected from a single germinal center, the rearranged heavy chain genes showed that several cells were clonally related, since they had the same VDJ joint. Genealogical trees were shown for the V genes expressed in two clones, indicating that successive somatic mutations must have occurred during several cycles of cell division. Similar isolation of single cells has also been used for the study of somatic mutation in germinal centers and marginal and mantle zones of human spleen and lymph nodes.[87,88]

Ideally, the process of somatic mutation would be followed in cultured cells to study the cellular requirements, molecular biology, and biochemistry of the reaction. Early on, an Abelson murine leukemia virus-transformed preB cell line, 18.81, was reported to undergo somatic mutation of endogenous Ig genes.[10] While preB cells do not normally mutate Ig genes, this cell line was also reported to show heavy chain class switching,[11] and thus to have traits of a mature B cell. Transfected Ig genes were also found to be mutated in the 18.81 cell line.[89] This same cell line was reported to undergo higher rates of somatic mutation after fusion to a myeloma line, NSO.[9] The assay used in these studies was to assess the reversion of a stop codon within the V, D, or C regions of a transfected heavy chain μ gene by immunofluorescence or an ELISA spot assay with anti-IgM. This may be a useful cell line to study somatic mutation. However, given that this cell line undergoes switch recombination, it may have the property of increased recombination efficiency rather than somatic mutation, which may cause reversion of the stop codon used as a mutation indicator. Furthermore, in some experiments with 18.81, a stop codon in the D region was found to be eliminated by a rearrangement and deletion mechanism, rather than by somatic hypermutation.[89]

Several attempts were made to culture fresh mouse B cells in the hope of establishing mutating cells in the normal physiological state in vitro. Microcultures of memory B cells defined by cell sorting for multiple surface markers were found to allow extensive cell proliferation after LPS stimulation.[90] The cells contained mutated Ig genes, but further somatic mutation in culture could not be demonstrated. Using a different stimulation protocol, including activated T cells, others were able to sustain somatic mutation in culture.[35,91,92] Kaellberg et al co-cultured spleen B cells from primed mice with helper T cells and anti-Ig at seven and ten days after immunization. After a four day culture, the cells were found to contain a higher overall frequency of somatic point mutations than at the start of the culture, suggesting that mutations had accumulated in vitro. To test the possibility that the more highly mutated cells survived preferentially, limiting dilution experiments were carried out allowing the outgrowth of separate B cell clones. While LPS-stimulated cultures showed no sequence diversity, clones that received helper T cells (Th2) plus anti-Ig showed a high frequency of mutations and intraclonal sequence diversity. Similar co-culture with activated T cells induced mutations in human tonsillar B cells.[92] These culture systems will be useful to investigate stimuli that maintain or perhaps induce the process of somatic mutation. At this time these systems may not be useful to study the mechanism of somatic mutation biochemically, because not enough cells can be obtained. Also, the introduction of mutation test substrates by transfection of freshly isolated B cells is currently not feasible.

Both these obstacles may possibly be overcome in a recently developed culture system using immortalized B cells.[93] The human Burkitt's lymphoma cell line BL2 was induced to mutate its expressed VH gene by activation with anti-IgM antibodies and co-cultivation with activated cloned T cells. The frequency of mutations after induction ranged between 3.5 and 27 x 10^{-4}. The cell line appeared not to mutate in the absence of stimuli, or if only the T cells or the anti-Ig were provided. These results look very promising. So far, they were only carried out with very small cell numbers (50-500 BL2 cells per well) and only with endogenous Ig genes. It is hoped that the system can be upscaled to allow biochemical analysis and, also, that transfected substrates can be introduced and will be mutated. Given that two different types of helper T cells, Th1 and Th2, were able to induce the mutations, it appears likely that the experimental setup may not be restricted to the T cells and perhaps the B cell line used in the original experiments.

Molecular Hallmarks of Somatic Mutation

In considering the possible molecular mechanisms responsible for somatic mutation it is important to know the specific features of this mutation process.

Mutation Rate

Compared with the overall spontaneous mutation rate, somatic hypermutation of Ig genes must operate at an extremely high rate. It has been estimated that the rate of introduction of base changes by this process is as high as one mutation per 1000 bp per cell generation,[94] i.e., about six orders of magnitude higher than the spontaneous mutation rate. This calculation is based on the time it takes to accumulate mutations in a given Ig gene, an assumed rapid cell generation time, and the average number of mutations introduced.

Types of Mutations

The mutations are mainly point mutations, with very rare deletions and even rarer insertions of mostly single nucleotides. Base transitions exceed transversions. Unfortunately, the latter finding may not allow to discrimination between different mechanisms, since many DNA polymerases, except AMV reverse transcriptase, have some bias for transitions.[95]

Strand Bias and A/T Bias

When inspecting the nucleotide changes due to somatic mutation, evidence for a strand bias was noted,[96-98] making it likely that the mutations are targeted to one of the DNA strands. Changes from adenine (A) are about twice as frequent as changes from thymine (T) when the top strand (the coding strand) is sequenced.[98] There often is also a differential for changes from G versus C; however, the difference is generally smaller and can go either way. The A>T preference was not seen for meiotically arisen mutations of pseudogenes.[96] The A>T preference has been interpreted as a strand bias, indicating that either of the two strands is preferentially, or exclusively, mutated. However, a strand bias alone would not be detectable, unless there is also an A/T bias.[99] If the A/T differential is caused by the process that causes the mutations, it can mean either of two mutually exclusive situations. If the top strand is where the mutations arise, A must be preferentially mutated, but if the bottom strand is the mutated one, T must be preferred.[99] However, one must consider that the bias may be introduced after the original mutagenic step. For example, mismatch repair may correct certain changes before the mutations are fixed (P. Gearhart, personal communication). In any case, one can hope that this signature combination of a strand bias and A/T bias will be a helpful guide to the unraveling of the mechanism of somatic mutation.

Hotspots

Certain nucleotide positions within a mutated sequence are found to be targeted more often than the rest of the sequence. For example, the G in the AGT serine codon at position 31 of Vκ-Ox is a hotspot of mutation.[100] This is even true with a passenger transgene that cannot be selected for antigen binding.[100] Hotspots of mutation are frequently observed and are at different positions in different target sequences, suggesting that the process is influenced by the primary sequence surrounding the hotspot.[86,98,101] Interestingly, V regions seem to be selected in evolution to be enriched in the hypervariable regions for particular mutational hotspots, such as the serine codons AGY (where Y is pyrimidine).[102] In the framework, on the other hand, the other serine codons, TCN (where N is any nucleotide), which are not mutational hotspots, dominate. Similarly, the observed targeting of somatic mutations in the sheep to the complementarity determining regions (CDR) of the V genes has been suggested to arise from sequence differences and special codon usage in the CDR versus the framework regions of the V genes.[30]

Substrates for Somatic Hypermutation

Besides the variable region, bacterial sequences (neomycin phosphotransferase, guanine phosphoribosyl transferase, chloramphenicol acetyl transferase), a globin gene, the Ig C region, and artificial substrates were also found to be somatically mutated in the context of an Ig transgene,[23,71,103,104] suggesting that the V region that is normally targeted for somatic mutation does not initiate the mutation process. However, the substrate can strongly influence the extent of somatic mutation. A short artificial sequence inserted into the V region of a kappa transgene was found to mutate about 10 times more frequently than the surrounding VJ region.[74] A different sequence from phage DNA has the opposite effect of inhibiting somatic mutation.[105] It will be very interesting to determine the molecular basis of these pronounced quantitative effects of the primary sequence on somatic mutation.

Both productively rearranged and nonproductively rearranged Ig genes can be mutated. Unrearranged Vκ genes have not been found to be mutated,[3,5] but unrearranged Vλ genes have.[106,107] This difference can presumably be attributed to the fact that in B cells the whole λ locus of about 200 kb[108,109] is in open chromatin and the unrearranged Vλ genes are transcribed,[110] whereas unrearranged Vκ genes are not transcribed.[111] Furthermore, at the heavy chain locus, DJ rearrangements that are allelic to a productive VDJ rearrangement can be somatically mutated, albeit at about a ten times lower frequency than VDJ genes.[112] The reason for this is unknown. There are transcriptional promoters upstream of the D genes. Perhaps they function at a reduced level for somatic mutation compared with the V region promoters (see below).

The 5' and 3' Boundaries of Somatic Mutation Tracts

The 5' boundary of the mutations is within the leader intron. Mutations upstream of the transcriptional start site are extremely rare.[113-116] Since it is likely that the mutation process is linked to transcription initiation (see below), the rare mutations upstream of the promoter are presumably due to the occasional transcription initiation from a cryptic upstream promoter.

There is no precise 3' end point of the mutation tract. However, the mutations only extend over about 1.5 kb with a slow decline of the frequency beyond the J region that is joined to the V gene[113,116-121] (Fig. 4.1, top). The 3' extent of the mutations depends on the position of the VJ joint in a particular gene: When a Vκ gene is rearranged to Jκ1, the mutations cease before Jκ5; none are found in the JC intron (Fig. 4.2). However, when Vκ is joined to Jκ5, point mutations can be found as far as 1 kb into the JC intron (Fig. 4.2).[63,119,120,122] In the λ locus, on the other hand, occasional mutations are seen in the C region.[118] The λ genes have a single J associated with each C region. It

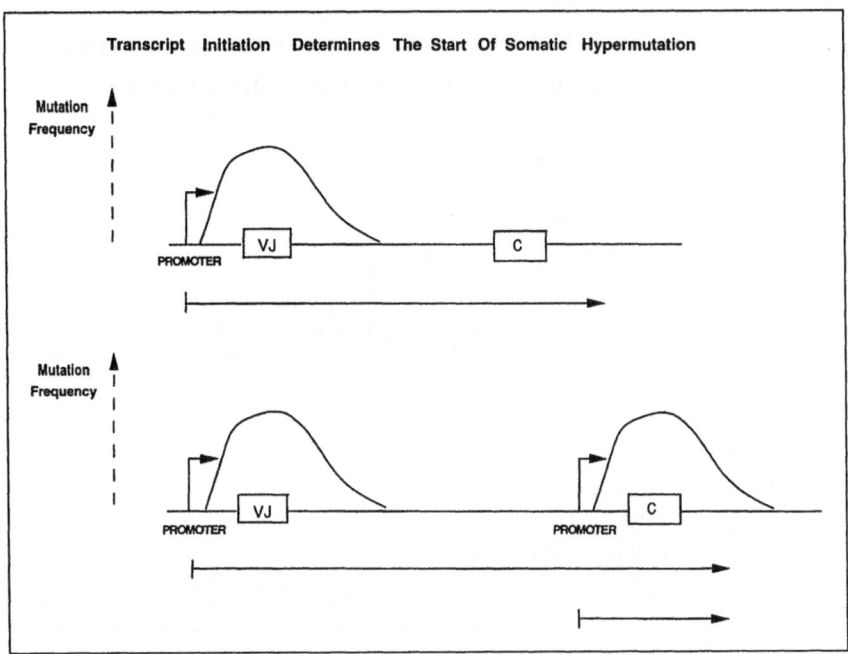

Fig. 4.1. Schematic distribution of somatic point mutations in Ig genes.

(Top) In wild type kappa genes, including transgenes, the mutated region spans from just 3' of the transcriptional promoter for about 1.5 kb to approximately 1 kb 3' of the VJ joint. The C region is not mutated.

(Bottom) In a kappa transgene with the promoter duplicated upstream of the C region, two mutated regions are observed, each starting 3' of the promoter. The region upstream of the second promoter is not mutated.

Transcripts for the two types of genes are indicated as horizontal bars with arrow heads. Modified from ref. 23.

is separated from the C region by an intron of only about 1.1 kb, compared to the 2.5 kb long intron between the most 3' J and the C region in the case of κ genes. Thus in both κ and λ genes the distance between the rearranged VJ region and the most 3' extent of the mutations is about the same.

The Role of the Promoter and Enhancers of Transcription

Transcription appears to be required for the somatic mutation process. Unrearranged Vκ genes that are not detectably transcribed[111] are not mutated,[3,5] but unrearranged Vλ genes that are transcribed[110] have been found to mutate.[106,107,118] This may suggest that the rate of transcription is tied to the rate of somatic mutation. However, no direct measurements of transcription rates within mutating B cells in the germinal centers have been reported. Curiously, the removal

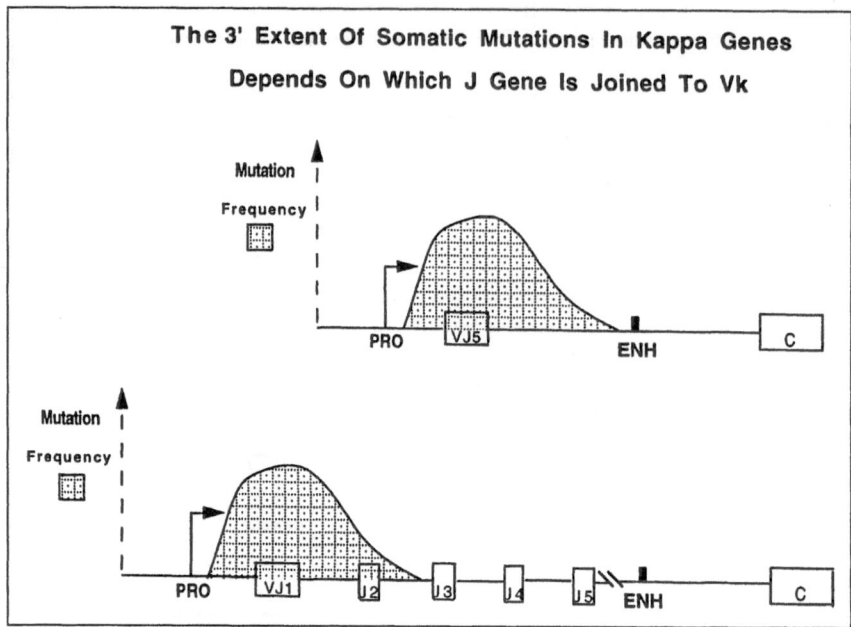

Fig. 4.2. Schematic distribution of somatic point mutations in kappa genes. (Top) When Vκ is rearranged to Jκ5, the mutations extend 3' into the middle of the JC intron. (Bottom) When Vκ is rearranged to Jκ1, the mutations reach only to about Jκ3. Modified from ref. 120.

of the κ intron enhancer (κi) eliminates measurable somatic mutation, even though the transgene is highly expressed in splenic hybridomas from these mice.[68] Also, a T cell receptor transgene under the control of the Ig heavy chain intron enhancer which is highly transcribed in total splenic B cells, is mutated at a very low rate.[64] In the case of the intron enhancer it now appears that the removal of the matrix attachment region (MAR) at the same time was responsible for the elimination of somatic mutation.[123] Perhaps Ig gene transcription is regulated differently in overall splenic B cells and in germinal centroblasts. It is conceivable that the MAR is essential for expression of the transgene in germinal center B cells, because the factors interacting with the 3' κ enhancer retained in the transgene may be at a low level at this stage of B cell life.[124] Likewise, the T cell receptor promoter may not cooperate with the heavy chain enhancer until the plasma cell stage.

Alternatively, there may be no direct relationship between transcription rate and mutation rate. While transcription seems to be required for somatic mutation, controls not directly linked to transcription may influence the mutation rate. In Ig transgenic mice different founder mice carrying the same transgene often show greatly different frequencies of somatic mutation, despite very similar transcript levels in splenic B cells. While, as mentioned above, there could be differential levels of expression in the germinal centers, it is also possible that elements present near the different insertion sites of the transgenes positively or negatively influence the mutation rate, without a major effect on transcription. This question remains a major challenge for investigation.

Another question is how transcriptional control may determine the specificity of the process. It has been shown that somatic mutation occurs at the normal frequency when the V promoter has been replaced by a β-globin promoter.[68] Also, replacement of the V promoter by the promoter of the B29 gene permits somatic mutation.[125] Furthermore, c-myc genes or Bcl-6 genes translocated into an Ig locus and transcribed under the control of a c-myc promoter or Bcl-6 promoter, respectively, are mutated.[126-129] This promoter promiscuity raises the question of whether the mutation process is really Ig gene specific (see below).

Ig Gene Specificity of Somatic Mutation

The only sequence that has so far been retained in all transgenic substrates that were found to be hypermutable is an Ig enhancer. In the κ locus, the intron enhancer (κi), but not the 3' enhancer (κ3'), appears to be indispensable.[71] In the λ locus, the λ enhancer suffices for somatic mutation.[73,104] However, this enhancer has little homology with the κi enhancer, except for certain E-boxes, and more resembles the κ3' enhancer.[130-132] Finally, heavy chain transgenes with only the heavy chain intron enhancer,[70,133] and a λ transgene with the same enhancer,[73,104] have been found to be mutable. Thus, the various Ig enhancers would be responsible for the Ig gene specificity of the mutation process. It is not clear how sequences as diverse as the κ intron enhancer and the λ enhancers, and which share binding sites for a variety of proteins involved in the regulation of other genes, can determine Ig gene specificity in the absence of the V promoter.

As an alternative, one needs to consider that many genes that are highly expressed in mutating B cells may be targets for the B cell specific mutator. This would present an additional hazard for the survival of mutating B cells. Any essential gene that is expressed in mutating B cells could be inactivated if both alleles suffered a deleterious mutation. Somatic hypermutation of certain non-Ig genes in tumor cells is compatible with vigorous cell growth. In Burkitt's lymphomas, *c-myc* genes translocated into the heavy chain locus are mutated.[126-128] This may be interpreted as Ig gene-like somatic hypermutation under the influence of the 3' heavy chain enhancer. Interestingly, the mutations show the same strand bias as somatic mutations in Ig genes,[128] and, in one study of blood lymphocytes with *c-myc* translocations, it appeared very likely that the mutations had occurred after the translocation event.[128] Furthermore, even *myc* genes that are not translocated into the heavy chain locus have been reported to be mutated.[134] In that study of several Burkitt's lymphoma cell lines with t(8;22) translocations, one of the *c-myc* genes was translocated to chromosome 22, into the λ locus. The other appeared not to be translocated, but nevertheless had undergone multiple alterations of restriction enzyme sites. If indeed this *c-myc* locus is not translocated, it would suggest that other expressed genes may be mutable in B cells undergoing somatic hypermutation. It is possible, of course, that the myc allele was indeed translocated into an Ig locus, but that the chromosomal break was outside the region where breaks normally occur on chromosome 8.

Similarly, *bcl-6* genes brought under the influence of the Ig heavy chain locus in diffuse large cell lymphomas and follicular lymphomas show high levels of somatic mutation.[129] In these cases too, nontranslocated *bcl-6* genes appear to be targeted at times. Again, these latter *bcl-6* genes may also have been translocated into an Ig locus without this being detectable. However, the possibility that many genes expressed in B cells undergoing somatic hypermutation can be mutated still needs to be investigated.

We have begun such an investigation, with the analysis of the transcription factor gene, Bcl-6, whose expression is upregulated in germinal centers.[135] "Memory" B cells from a normal individual were used whose Ig heavy chain genes were highly mutated.[146] Surprisingly, Bcl-6 genes were also clearly mutated in these cells, although at a 10 to 100 times lower frequency than the heavy chain genes. However, Bcl-6 genes were not mutated in "naive" B cells. A number of other genes were also not mutated in the "memory" B cells.

In summary, it now appears that the somatic mutation process is generally restricted to Ig genes, but that Bcl-6 (and perhaps other genes?) may share control sequences that specifically target a postulated mutator factor (see below). The basis for this Ig specificity needs to be determined.

A Model for the Mechanism of Somatic Hypermutation

Several models have been proposed to explain the molecular basis of somatic hypermutation: (1) site specific nicking of the DNA followed by repair via an error-prone DNA polymerase;[122,136] (2) homologous recombination of reverse transcribed Ig mRNA;[137] (3) somatic gene conversion;[138] (4) V region DNA replication independent of chromosomal DNA replication;[139] and (5) alteration of the fidelity of DNA replication.[140] These models have been reviewed recently elsewhere.[53]

A new model proposed by our laboratory is based on the following observations and considerations. Normally, κ genes are mutable over about a 1.5 kb region starting in the leader intron.[116,119,120] In a κ gene with a VJ5 rearrangement the frequency of point mutations peaks over VJ and declines to zero towards the middle of the JC intron (Fig. 4.2).[119,120] The C region is normally not mutated. The same distribution is seen in κ transgenes that mimic the structure of endogenous κ genes.[63] However, in a κ transgene in which the transcriptional promoter has been duplicated upstream of the C region, two waves of somatic hypermutation are observed, one over the VJ region and another one over the C region[23] (Fig. 4.1, bottom). In this situation, two types of transcripts are produced in about equal quantities. One type initiates from the upstream promoter and ends 3' of C, another initiates from the internal promoter. Thus, it appears that initiation of transcription directly upstream of the C region parallels initiation of hypermutation in that region.

These results suggest that in the somatic hypermutation process a mutator factor is transported into the gene from the transcriptional promoter. The findings led to the formulation of a new model of somatic hypermutation. The model postulates that a mutator factor (MuF) binds to the RNA polymerase at the promoter, travels with the polymerase during the elongation phase of transcription, and induces pausing of the polymerase.[23,53,99,141] The pausing was proposed to lead to a stable polymerase/MuF complex which is recognized as a DNA lesion by a DNA repair system, such as transcription coupled nucleotide excision repair (NER). During NER the single stranded DNA with the bound complex was proposed to be excised and resynthesized with the occasional introduction of errors. These errors were postulated to represent the somatic point mutations.

To test the NER aspect of the new somatic mutation model, we have investigated B lymphocytes from patients and mice with severe defects in NER.[83,142] Epstein Barr Virus-transformed B lymphocytes

from patients suffering from various forms of xeroderma pigmentosum (XP-B, XP-D, and XP-V) or of Cockayne syndrome (CS-A)
showed high levels of somatic mutation in the V and J regions of
their expressed heavy and light chain genes.[83] The C regions were
not mutated, suggesting that the V and J region mutations were indeed due to somatic hypermutation. In further support of somatic
hypermutation, V genes associated with switched CH genes, were
more highly mutated than those associated with Cμ genes and the
ratios of replacement to silent mutations were greater in the
complementarity determining regions than in the framework regions
of the V genes. The four patients were chosen because they had signs
of transcriptional defects, suggesting special deficiencies in transcription coupled DNA repair. It appeared, therefore, that transcription-
coupled NER was not required for somatic hypermutation. Also, an
XP-D defect without transcription defects does not interfere with
somatic mutation.[80] Finally, we investigated mice with a knockout
of yet another NER gene, XP-C, that has been implicated in the repair of the nontranscribed DNA strand.[142] These mice as well showed
normal levels of somatic hypermutation.

Recently, DNA mismatch repair (MMR) deficiency was shown
to eliminate the preferential repair of the transcribed DNA strand.[143]
It was therefore possible that MMR was required for somatic
hypermutation. Mice with defects in the mammalian homologues
of the *E. coli* MMR genes Mut L (MLH1, PMS2) have been investigated for their ability to somatically mutate Ig genes (Gearhart et al,
personal communication; N. Kim, J. Lo, and U. Storb, unpublished).
All appeared capable of high levels of somatic hypermutation. There
was some suggestion, however, that the distribution and fine patterns of the mutations may be altered (P. Gearhart, personal communication). This question is under further investigation.

These data combined suggested that general mechanisms of
DNA repair have not been usurped for the creation of occasional
errors in the process of somatic hypermutation of Ig genes. However, recent experiments have given new clues as to how the mutations may be introduced. We have created an Ig κ transgene that carries, inserted into the V region, an artificial mutation substrate.[74]
The substrate consists of alternating *Eco*RV and *Pvu*II restriction
enzyme sites (EPS). Mutations in the EPS are easily assayed by DNA
amplification and restriction enzyme analysis. The products can be
electrophoretically separated, resulting in a ladder of small fragments

when no mutations have occurred. When a mutation exists in the EPS, two small fragments disappear and a larger one is created. The EPS transgene is thus a useful substrate for a quick mutation assay.

When transgene copies with EPS mutations were sequenced, including the flanking V and J regions, we found that the EPS is on average mutated 7x more frequently than the flanks.[144] Moreover, within the EPS a hierarchy of mutability exists. Certain nucleotide combinations are mutated much more frequently than others in every restriction site that contains the respective nucleotides. For example, GC and TA appear hypermutable, whereas TG and GA are disfavored. The preferred targets turned out to be components of hotspots observed in a number of studies of somatic hypermutations.[98,101,145] In fact, comparing the EPS sequence with the sequences analyzed in a large survey of somatically mutated Ig genes[98] showed a rather striking correlation of hyper- and hypo-mutability within the EPS with the di- and tri-nucleotides that were favored or disfavored for somatic mutation of normal Ig genes.

This alone did not explain, however, why the EPS was hypermutable compared with the V and J flanks. These flanks also contained similar accumulations of "hotspots". Unlike the flanks, though, the EPS sequence, because of its repeated restriction sites, can form a series of stem-loop structures with very high stability of the stems. Such secondary structures could form either on the DNA or the RNA level. We have proposed that they play a role in somatic mutation (see below).

Two major rules for somatic hypermutation were suggested by the results with the EPS transgene. First, the overall region where point mutations occur may be determined by the secondary structure of the nascent RNA. Second, the exact nucleotide changes depend on the nucleotide preferences of a mutator factor that is specific for the somatic hypermutation of Ig genes. The findings with the EPS transgene have led us to modify the original transcription-based model[23] by postulating that the pausing is the natural pausing that occurs during normal transcription (Fig. 4.3).[144] Furthermore, the role of a mutator factor (MuF) is to directly interact with DNA near the pause site to cause the mutations. For example, MuF could be a nuclease that nicks DNA, which is followed by excision of a short patch of single stranded DNA by general exonucleases. Repolymerization of the patch could be error prone due to the influence of the MuF.

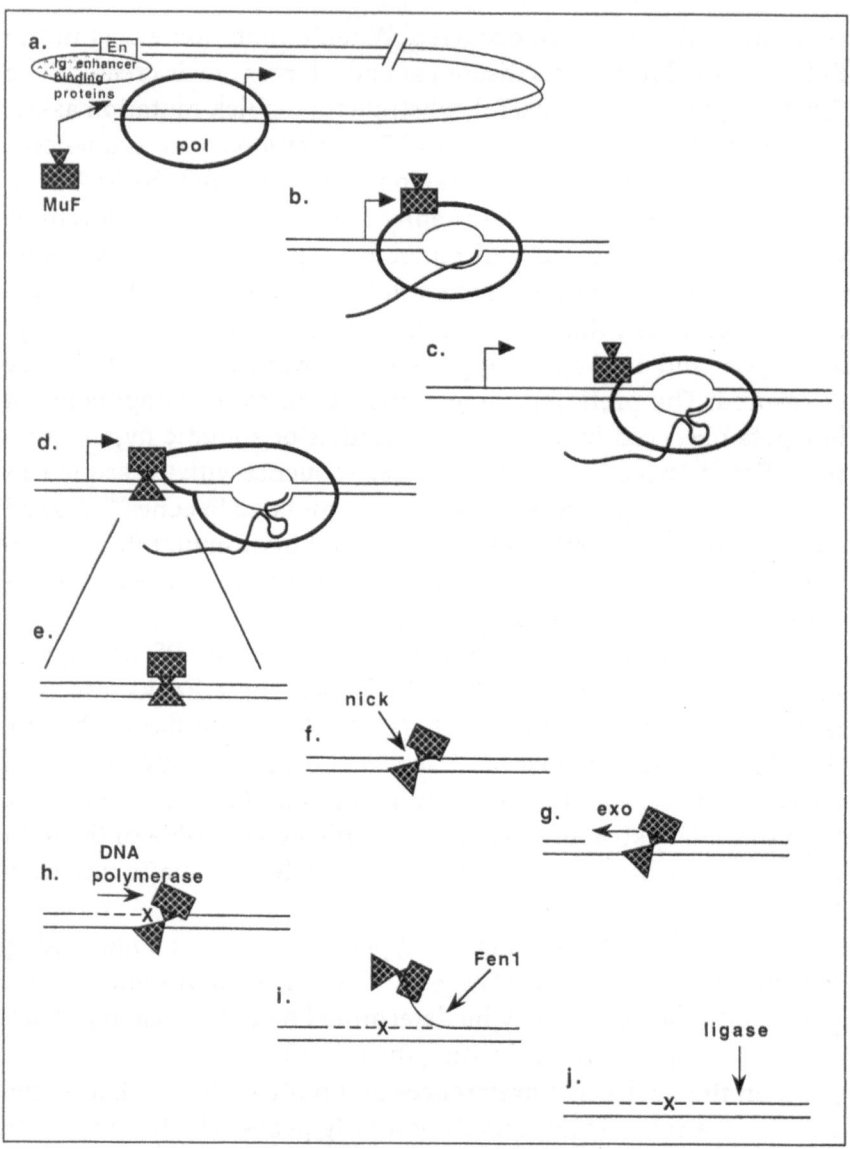

Fig. 4.3. Model of somatic hypermutation of Ig genes. En, Ig enhancer; MuF, mutator factor; pol, RNA polymerase II; exo, DNA exonuclease; x, point mutation; Fen1, flap endonuclease 1. Modified from ref. 144.

The proposed model is summarized in Fig. 4.3.

a. MuF associates with RNA polymerase II (pol) that has interacted with Ig enhancer binding proteins during initiation of transcription at an Ig gene promoter.

b. MuF travels with the polymerase during transcript elongation.

c. If the RNA polymerase encounters a block to transcription (e.g., a hairpin in the nascent transcript), it pauses.

d. Perhaps under the influence of elongation factors, the conformation of the pausing polymerase changes, resulting in the transfer of MuF to the DNA.

e. MuF binds to double stranded DNA upstream of the polymerase.

How the MuF then induces mutations is rather speculative, but may proceed in the following way:

f. MuF causes a nick in the nontranscribed strand; it remains associated with the newly created 5' end and also with one or several nucleotides on the transcribed strand.

g. Because of the bound MuF, the single stranded ends cannot be ligated. Exonuclease trims back the single stranded 3' end.

h. A DNA polymerase fills in the gap creating mutation(s) (x) opposite the MuF associated base(s).

i. The DNA polymerase continues past the MuF bound residues, creating a 5' flap of the nontranscribed strand. The flap is cut by endonuclease Fen1 (DNase IV).

j. The free DNA ends are ligated. MuF has been removed with the flap.

The postulated MuF would only be produced during the short window in the life of the B cell when somatic hypermutation is ongoing. It can only bind to an initiating RNA polymerase II (pol) that has interacted with transcription factors bound to the Ig enhancer, not randomly with pol or DNA. This property of MuF, coupled with a high chance for pausing within the first 1 kb or so of transcribed DNA, explains the extent of somatic hypermutations over only 1-2 kb of the 5' region of the Ig gene with sparing of the 3' region (reviewed in ref. 53).

This novel model of somatic hypermutation of Ig genes has several aspects that can be further tested by studying the proteins associated with the EPS transgene during somatic mutation and creating modifications of the EPS transgene.

Conclusions

The process of somatic hypermutation of Ig genes takes place exclusively in germinal center B lymphocytes. Interaction with T cells and antigen drives the B cell into the cell cycle and induces a mutator factor whose identity and properties are unknown. The mutation process requires transcription, is initiated around the transcriptional promoter, and extends for approximately 1.5 kb into the targeted gene. Targeting appears to be Ig gene specific. The mechanism of inducing the mutations is unknown. Nucleotide excision repair and DNA mismatch repair do not seem to be required. As shown with an artificial substrate, the mutations appear to be dependent on both the secondary structure and the primary sequence of the target genes. Novel experimental systems to study somatic mutation have been developed and are expected to lead to the further understanding of this fascinating phenomenon.

Acknowledgments.

I am very grateful to S. Denepoux, A. Peters and N. Kim for critical reading of the manuscript and numerous suggestions, and to C. Berek, R. Dalla-Favera, S. Denepoux, M. Flajnik, N. Green, P. Gearhart, G. Kelsoe, S. Lebeque, T. Leanderson, C. Milstein, B. Rogerson, and M. Scharff for stimulating discussions and/or providing preprints and reprints of their research. The work cited from my laboratory was supported by NIH grant GM38649.

References

1. Weigert M, Cesari IM, Yonkovich SJ, Cohn M. Variability in the lambda light chain sequences of mouse antibody. Nature 1970; 228:1045-1047.
2. Bernard O, Hozumi N, Tonegawa S. Sequences of mouse immunoglobulin light chain genes before and after somatic changes. Cell 1978; 15:1133-1144.
3. Selsing E, Storb U. Somatic mutation of immunoglobulin light-chain variable-region genes. Cell 1981; 25:47-58.
4. Crews S, Griffin J, Huang H, Calame K, Hood L. A single VH gene segment encodes the immune response to phosphorylcholine: Somatic mutation is correlated with the class of the antibody. Cell 1981; 25:59-66.
5. Gorski J, Rollini P, Mach B. Somatic mutations of immunoglobulin variable genes are restricted to the rearranged V gene. Science 1983; 220:1179-1181.
6. Engler P, Klotz E, Storb U. N region diversity of a transgenic substrate in fetal and adult lymphoid cells. J Exp Med 1992; 176: 1399-1404.

7. Lewis SM. The mechanism of V(D)J joining: Lessons from molecular, immunological and comparative analyses. Adv Immunol 1994; 56:27-150.

8. Bogue M, Roth DB. Mechanism of V(D)J recombination. Curr Op Immunol 1996; 8:175-180.

9. Green N, Rabinowitz J, Zhu M, Kobrin B, Scharff M. Immunoglobulin variable region hypermutation in hybrids derived from a pre-B- and a myeloma cell line. Proc Natl Acad Sci USA 1995; 92:6304-6308.

10. Wabl M, Burrows P, von Gabain A, Steinberg C. Hypermutation at the immunoglobulin heavy chain locus in a pre-B-cell line. Proc Natl Acad Sci USA 1985; 82:479-482.

11. Jaeck H-M, McDowell M, Steinberg C, Wabl M. Looping out an deletion mechanism for the immunoglobulin heavy-chain class switch. Proc Natl Acad Sci USA 1988; 85:1581-1585.

12. Manser T, Gefter M. The molecular evolution of the immune response: Idiotype specific suppression indicates that B cells express germline encoded V genes prior to antigenic stimulation. Eur J Immunol 1986; 16:1439-1444.

13. Klein U, Kueppers U, Rajewsky K. Human IgM[+]IgD[+] B cells, the major B cell subset in the peripheral blood, express V-kappa genes with no or little somatic mutation throughout life. Eur J Immunol 1993; 23:3272-3277.

14. Liu Y-J, Zhang J, Lane P, Chan E, MacLennan I. Sites of specific B cell activation in primary and secondary responses to T cell-dependent and T cell-independent antigens. Eur J Immunol 1991; 21:2951-2962.

15. Kimoto H, Nagaoka H, Adachi Y et al. Accumulation of somatic hypermutation and antigen-driven selection in rapidly cycling surface Ig[+] germinal center (GC) B cells which occupy GC at a high frequency during the primary anti-hapten response in mice. Eur J Immunol 1997; 27:268-279.

16. Vonderheide R, Hunt S. Comparison of IgD[+] and IgD[-] thoracic duct B lymphocytes as germinal center precursor cells in the rat. Internatl Immunol 1991; 3:1273-1281.

17. Liu Y-J, de Bouteiller O, Arpin C et al. Normal human IgD+IgM-germinal center B cells can express up to 80 mutations in the variabale region of their IgD transcripts. Immunity 1996; 4:603-613.

18. Liu Y, Malisan F, de Bouteiller O et al. Within germinal centers, isotype switching of immunoglobulin genes occurs after the onset of somatic mutation. Immunity 1996; 4:241-250.

19. Shan H, Schlomchik M, Weigert M. Heavy-chain class switch does not terminate somatic mutation. J Exp Med 1990; 172:531-536.

20. Gearhart P, Johnson N, Douglas R, Hood L. IgG antibodies to phosphorylcholine exhibit more diversity than their IgM counterparts. Nature 1981; 291:29-34.

21. Klein U, Kueppers R, Rajewsky K. Variable region gene analysis of B cell subsets derived from a 4-year-old child: Somatically mutated memory B cells accumulate in the peripheral blood already at a young age. J Exp Med 1994; 180:1383-1393.

22. Pascual V, Liu Y, Magalski A, de Bouteiller O, Banchereau J, Capra D. Analysis of somatic mutation in five B cell subsets of human tonsil. J Exp Med 1994; 180:329-339.
23. Peters A, Storb U. Somatic hypermutation of immunoglobulin genes is linked to transcription initiation. Immunity 1996; 4:57-65.
24. MacLennan I, Liu Y, Oldfield S, Zhang J, Lane P. The evolution of B cell clones. Curr Top Microbiol Immunol 1990; 159:37-60.
25. Berek C, Berger A, Apel M. Maturation of the immune response in germinal centers. Cell 1991; 67:1121-1129.
26. Tew JG, Kosko MH, Burton GF, Szakal AK. Follicular dendritic cells as accessory cells. Immunol Rev 1990; 117:185-211.
27. Kelsoe G. In situ studies of germinal center reaction. Adv Immunol 1995; 60:267-288.
28. Jacob J, Kelsoe G, K. R, Weiss U. Intraclonal generation of antibody mutants in germinal centers. Nature 1991; 354:389-392.
29. Jacob J, Kassir R, Kelsoe G. In situ studies of the primary immune response to (4-hydroxy-3-nitrophenyl)acetyl. I. The architecture and dynamics of responding cell polulations. J Exp Med 1991; 173: 1165-1175.
30. Reynaud C, Garcia C, Hein W, Weill J. Hypermutation generating the sheep immunoglobulin repertoire is an antigen-independent process. Cell 1995; 80:115-125.
31. Carter M, Li S, Wilkinson M. A splicing-dependent regulatory mechanism that detects translation signals. EMBO J 1996; 15: 5965-5975.
32. Shlomchik MJ, Marshak-Rothstein A, Wofowicz CB, Rothstein TL, Weigert M. The role of clonal selection and somatic mutation in autoimmunity. Nature 1987; 328:805-811.
33. Siekevitz M, Kocks C, Rajewsky K, Dildrop R. Analysis of somatic mutation and class switching in naive and memory B cells generating adoptive primary and secondary responses. Cell 1987; 48:757-770.
34. Griffiths G, Berek C, Kaartinen M, Milstein C. Somatic mutation and the maturation of immune reponse to 2-phenyl oxazolone. Nature 1984; 312:271-275.
35. Decker D, Linton P, Zaharevitz S, Biery M, Gingeras T, Klinman N. Defining subsets of naive and memory B cells based on the ability of their progeny to somatically mutate in vitro. Immunity 1995; 2:195-203.
36. Bothwell A, Tao W, Blier P. A model for the development of somatic variants in memory B cells. In: Steele EJ, ed. Somatic Hypermutation in V Regions. Boca Raton, FL: CRC Press, 1991:55-67.
37. Kipps TJ. The CD5 B cell. Adv Immunol 1988; 47:117-185.
38. Ebeling S, Schutte M, Logtenberg T. Peripheral human CD5$^+$ and CD5- B cells may express somatically mutated VH5- and VH6-encoded IgM receptors. J Immunol 1993; 151:6891-6899.
39. Hashimoto S, Dono M, Wakai M et al. Somatic diversification and selection of Ig heavy and light chain variable region genes in IgG$^+$ CD5$^+$ chronic lymphocytic leukemia B cells. J Exp Med 1995; 181:1507-1517.

40. Zheng B, Xue W, Kelsoe G. Locus-specific somatic hypermutation in germinal centre T cells. Nature 1994; 372:556-559.
41. Vitetta E, Berton M, Burger C, Kepron M, Lee W, Yin X-M. Memory B and T cells. Ann. Rev Immunol 1991; 9:193-217.
42. McHeyzer-Williams M, Davis M. Antigen-specific development of primary and memory T cells in vivo. Science 1995; 268:106-110.
43. Vescio R, Cao J, Hong C et al. Myeloma Ig heavy chain V region sequences reveal prior antigenic selection and marked somatic mutation but no intraclonal diversity. J Immunol 1995; 155:2487-2497.
44. Klein U, Klein G, Ehlin-Hendricksson B, Rajewsky K, Kueppers R. Burkitt's lymphoma is a malignancy of mature B cells expressing somatically mutated V region genes. Mol Med 1995; 1:495-505.
45. Kueppers R, Hajadi M, Plank L, Rajewsky K, Hansmann M-L. Molecular Ig gene analysis reveals that monocytoid B cell lymphoma is a malignancy of mature B cells carrying somatically mutated V region genes and suggests that rearrangment of the kappa-deleting element (resulting in the deletion of the Ig kappa enhancers) abolishes somatic hypermutation in the human. Eur J Immunol 1996; 26:1794-1800.
46. Levy R, Levy S, Cleary M et al. Somatic mutation in human B-cell tumors. Imm Rev 1987; 96:43-58.
47. Bahler D, Levy R. Clonal evolution of a follicular lymphoma: Evidence for antigen selection. Proc Natl Acad Sci USA 1992; 89: 6770-6774.
48. Cuisinier A, Gauthier L, Boubli L, Fougerau M, Tonnelle C. Mechanisms that generate human Ig diversity operate from the 8th week of gestation in fetal liver. Eur J Immunol 1993; 23:110-118.
49. Reynaud C, Mackay C, Mueller R, Weill J. Somatic generation of diversity in a mammalian primary lymphoid organ: The sheep ileal Peyer's patches. Cell 1991; 64:995-1005.
50. Maybaum T, Reynolds J. B cells selected for apoptosis in the sheep ileal Peyer's patch have enhanced mutational diversity in the Ig V-lambda licht chain. J Immunol 1996; 157:1474-1484.
51. Becker R, Knight K. Somatic diversification of Ig heavy chain VDJ genes: Evidence for somatic gene conversion in rabbits. Cell 1990; 63:987-997.
52. Reynaud C, Anquez V, Grimal H, Weill J-C. A hyperconversion mechanism generates the chicken light chain preimmune repertoire. Cell 1987; 48:379-388.
53. Storb U. The molecular basis of somatic hypermutation of immunoglobulin genes. Curr Op Immunol 1996; 8:206-214.
54. Lanning DK, Knight K. Somatic hypermutation: Mutations 3' of rabbit VDJ h-chain genes. J Immunol 1997;submitted.
55. Parvari R, Ziv E, Lantner F, Heller D, Schechter I. Somatic diversification of chicken immunoglobulin light chains by point mutations. Proc Natl Acad Sci USA 1990; 87:3072-3076.

56. Wilson M, Hsu E, Marcuz A, Courtet M, Du Pasquier L, Steinberg C. What limits affinity maturation of antibodies in Xenopus—the rate of somatic mutation or the ability to select mutants? EMBO J 1992; 11:4337-4347.
57. Greenberg A, Avilla D, Hughes M, Hughes A, McKinney E, Flajnik M. A new antigen receptor gene family that undergoes rearrangement and extensive somatic diversification in sharks. Nature 1995; 374:168-173.
58. Gearhart P. Somatic muttion and affinity maturation. In: Paul WE, ed. Fundamental Immunology. 3rd ed. New York: Raven Press, Ltd., 1993:865-885.
59. Levy N, Malipiero U, Lebeque S, Gearhart P. Early onset of somatic mutation in immunoglobulin VH genes during the primary immune response. J Exp Med 1989; 169:2007-2019.
60. Cumano A, Rajewsky K. Clonal recruitment and somatic mutation in the generation of immunological memory to the hapten NP. EMBO J 1986; 5:2459-2468.
61. Wysocki L, Manser T, Gefter M. Somatic evolution of variable region structures during an immune response. Proc Natl Acad Sci USA 1986; 83:1847-1851.
62. O'Brien R, Brinster R, Storb U. Somatic hypermutation of an immunoglobulin transgene in κ transgenic mice. Nature 1987; 326:405-409.
63. Hackett J, Rogerson B, O'Brien R, Storb U. Analysis of somatic mutations in κ transgenes. J Exp Med 1990; 172:131-137.
64. Hackett J, Stebbins C, Rogerson B, Davis M, Storb U. Analysis of a T cell receptor gene as a target of the somatic hypermutation mechanism. J Exp Med 1992; 176:225-231.
65. Sharpe M, Milstein C, Jarvis J, Neuberger M. Somatic hypermutation of immunoglobulin κ may depend on sequences 3' of Cκ and occurs on passenger transgenes. EMBO J 1991; 10:2139-2145.
66. Sharpe M, Neuberger M, Pannell R, Surami A, Milstein C. Lack of somatic mutation in a κ light chain transgene. Eur J Immunol 1990; 20:1379-1385.
67. Carmack C, Camper S, Mackle J, Gerhard W, Weigert M. Influence of a Vκ8 L chain transgene on endogenous rearrangements and the immune response to the HA(SB) determinant of influenza virus. J Immunol 1991; 147:2024-2033.
68. Betz A, Milstein C, Gonzalez-Fernandes R, Pannell R, Larson T, Neuberger M. Elements regulating somatic hypermutation of an immunoglobulin K gene: Critical role for the intron enhancer/matrix attachment region. Cell 1994; 77:239-248.
69. Giusti A, Manser T. Hypermutation is observed only in antibody H chain V region transgenes that have recombined with endogenous immunoglobulin H DNA: Implications for the location of cis-acting elements required for somatic mutation. J Exp Med 1993; 177:797-809.
70. Sohn J, Gerstein R, Hsieh C, Lemer M, Selsing E. Somatic hypermutation of an immunoglobulin μ heavy chain transgene. J Exp Med 1993; 177:493-504.

71. Yelamos J, Klix N, Goyenechea B et al. Targeting of non-Ig sequences in place of the V segment by somatic hypermutation. Nature 1995; 376:225-229.
72. Johnston J, Ihyer S, Smitch R et al. Analysis of hypermutation in immunoglobulin heavy chain passenger transgenes. Eur J Immunol 1996; 26:1058-1062.
73. Klotz E, Storb U. Somatic hypermutation of a lambda-2 transgene under the control of the lambda enhancer or the heavy chain intron enhancer. J Immunol 1996; 157:4458-4463.
74. Klotz E, Hackett JJ, Storb U. Somatic hypermutation of an artificial test substrate within an Ig kappa transgene. Immunol 1998; (in press).
75. Carter M, Li S, Wilkinson M. A splicing-dependent regulatory mechanism that detects translation signals. EMBO J 1996; 15: 5965-5975.
76. Rose M, Birbeck S, Wallis V, Forrester J, Davies A. Peanut lectin binding properties of germinal centers of mouse lymphoid tissue. Nature 1980; 284:364-366.
77. Apel M, Berek C. Somatic mutations in antibodies expressed by germinal center B cells early after primary immunization. Internatl Immunol 1990; 2:813-819.
78. Gonzales-Fernandes A, Milstein C. Analysis of somatic hypermutation in mouse Peyer's patches using immunoglobulin κ light-chain transgenes. Proc Natl Acad Sci USA 1993; 90:9862-9866.
79. Gonzalez-Fernandez A, Gilmore D, Milstein C. Age-related decrease in the proportion of germinal center B cells from mouse Peyer's patches is accompanied by an accumulation of somatic mutations in their immunoglobulin genes. Eur J Immunol 1994; 24:2918-2921.
80. Wagner S, Elvin J, Norris P, McGregor J, Neuberger M. Somatic hypermutation of Ig genes in patients with xeroderma pigmentosum (XP-D). Internatl Immunol 1996; 8:701-705.
81. Chu Y, Marin E, Fuleihan R et al. Somatic mutation of human Ig V genes in the X-linked hyperIgM syndrome. J Clin Invest 1995; 95:1389-1393.
82. Timmers E, Hermans M, Kraakman M, Hendriks R, Schuurman R. Diversity of immunoglobulin kappa light chain gene rearrangements and evidence for somatic mutation in V-kappa IV family gene segments in X-linked agammaglobulenemia. Eur J Immunol 1993; 23:619-624.
83. Kim N, Kage K, Matsuda F, Lefranc M-P, Storb U. B lymphocytes of xeroderma pigmentosum or Cockayne syndrome patients with inherited defects in nucleotide excision repair are fully capable of somatic hypermutation of immunoglobulin genes. J Exp Med 1997; 186:413-419.
84. Varade W, Insel R. Isolation of germinal centerlike events from human spleen RNA. J Clin Invest 1993; 91:1838-1842.
85. Insel R, Varade W, Marin E. Human splenic IgM Ig transcripts are mutated at high frequency. Mol Immunol 1994:383-392.

86. Insel R, Varade W. Bias in somatic hypermutation of human VH genes. Internatl Immunol 1994; 6:1437-1443.
87. Kueppers R, Zhao M, Hansmann M, Rajewsky K. Tracing B cell development in human germinal centres by molecular analysis of single cells picked from histological sections. EMBO J 1993:4955-4967.
88. Dunn-Walters D, Isaacson P, Spencer J. Analysis of mutations in Ig heavy chain V region genes of microdissected marginal zone (MGZ) B cells suggests that the MGZ of human spleen is a reservoir of memory b cells. J Exp Med 1995; 182:559-566.
89. Bachl J, Wabl M. An immunoglobulin mutator that targets G/C base pairs. Proc Natl Acd Sci USA 1996; 93:851-855.
90. McHeyzer-Williams M, Nossal G, Lalor P. Molecular characterization of single memory B cells. Nature 1991; 350:502-505.
91. Kaellberg E, Jainandunsing S, Gray D, Leanderson T. Somatic mutation of immunoglobulin V genes in vitro. Science 1996; 271: 1285-1289.
92. Razanajaona D, Denepoux S, Blanchard D et al. In vitro triggering of somatic mutation in human naive B cells. J Immunol 1997; 159:3347-3353.
93. Denepoux S, Razanajaona D, Blanchard D et al. Induction of somatic mutation in a human B cell line in vitro. Immunity 1997; 6:35-46.
94. McKean D, Huppi K, Bell M, Staudt L, Gerhard W, Weigert M. Generation of antibody diversity in the immune response of BALB/c mice to influenza virus hemagglutination. Proc Natl Acad Sci 1984; 81:3180-3184.
95. Kunkel T. Hypermutation during DNA synthesis in vitro. In: Steele EJ, ed. Somatic Hypermutation in V-Regions. Boca Raton Ann Arbor, Boston: CRC Press, 1991:159-178.
96. Golding G, Gearhart P, Glockman B. Patterns of somatic mutation in immunoglobulin variable genes. Genetics 1987; 115:169-176.
97. Neuberger MS, Milstein C. Somatic hypermutation. Curr Op Imm 1995; 7:248-254.
98. Smith D, Creadon G, Jena P, Portanova J, Kotzin B, Wysocki L. Di- and trinucleotide target preferences of somatic mutagenesis in normal and autoreactive B cells. J Immunol 1996; 156:2642-2652.
99. Storb U, Peters A, Klotz E et al. Somatic hypermutation of immunoglobulin genes is linked to transcription. Curr. Topics Microbiol Immunol 1998; 229:11-19.
100. Betz AG, Rada C, R. P, Milstein C, Neuberger M. Passenger transgenes reveal intrinsic specificity of the antibody hypermutation mechanism: Clustering, polarity and specific hotspots. Proc Natl Acad Sci 1993; 90:2385-2388.
101. Doerner T, Brezinschek H-P, Brezinschek R, Foster S, Domiati-Saad R, Lipsky P. Analysis of the frequency and pattern of somatic mutations ithin nonproductively rearranged human variable heavy chain genes. J Immunol 1997; 158:2779-2789.

102. Wagner S, Milstein C, Neuberger M. Codon bias targets mutation. Nature 1995; 376:732.
103. Azuma T, Motoyama N, Fields L, Loh D. Mutations of the chloramphenicol acetyl transferase transgene driven by the immunoglobulin promoter and intron enhancer. Internatl Immunol 1993; 5:121-130.
104. Klotz E. An analysis of molcular requirements for somatic hypermutation using immunoglobulin light chain transgenes [Ph.D.]. Chicago: University of Chicago, 1997:132.
105. Winter D, Sattar N, Mai J-J, Gearhart P. Insertion of 2 kb of bacteriophage DNA between an immunoglobulin promoter and leader exon stops somatic hypermutation in a kappa transgene. Mol Immunol 1997; 34:359-366.
106. Weiss S, Wu G. Somatic point mutations in unrearranged immunoglobulin gene segments encoding the variable region of lambda light chains. EMBO J 1987; 6:927-932.
107. Motoyama N, Miwa T, Suzuki Y, Okada H, Azuma T. Comparison of somatic mutation frequency among immunoglobulin genes. J Exp Med 1994; 179:395-403.
108. Storb U, Haasch D, Arp B, Sanchez P, Cazenave P, Miller J. Physical linkage of mouse λ genes by pulsed-field gel electrophoresis suggests that the rearrangement process favors proximate target sequences. Mol Cell Biol 1989; 9:711-718.
109. Carson S, Wu G. A linkage map of the mouse immunoglobulin λ light chain locus. Immunogenetics 1989; 29:173-179.
110. Picard D, Schaffner W. Unrearranged immunoglobulin lambda variable region is transcribed in kappa-producing myelomas. EMBO J 1984; 3:3031-3035.
111. Mather E, Perry R. Transcriptional regulation of immunoglobulin V genes. Nucleic Acids Res 1981; 9:6855-6867.
112. Roes J, Hueppi K, Rajewsky K, Sablitzky F. V gene rearrangement is required to fully activate the hypermutation mechanism in B cells. J Immunol 1989; 142:1022-1026.
113. Both G, Taylor L, J. P, Steele E. Distribution of mutations around rearranged heavy-chain antibody variable-region genes. Mol Cell Biol 1990; 10:5187-5196.
114. Gearhart PJ, Levy NS. Kinetics and molecular model for somatic mutation in immunoglobulin variable genes. In: Steele EJ, ed. Somatic hypermutation in V-regions. Boca Raton: CRC Press, 1991: 29-39.
115. Rada C, Gonzalez-Fernandez A, Jarvis JM, Milstein C. The 5' boundary of somatic hypermutation in a Vk gene is in the leader intron. Eur J immunol 1994; 24:1453-1457.
116. Rogerson B. Mapping the upstream boundary of somatic mutations in rearranged immunoglobulin transgenes and endogenous genes. Mol Immunol 1994; 31:83-98.
117. Lebecque S, Gearhart P. Boundaries of somatic mutation in rearranged immunoglobulin genes: 5' boundary is near the promoter, and 3' boundary is ~1kb from V(D)J gene. J Exp Med 1990; 172:1717-1727.

118. Motoyama N, Okada H, Azuma T. Somatic mutation in constant region of mouse λ1 light chains. Proc Natl Acad Sci USA 1991; 88:7933-7937.
119. Weber JS, Berry J, Litwin S, Claflin JL. Somatic hypermutation of the JC intron is markedly reduced in unrearranged κ and H alleles and is unevenly distributed in rearranged alleles. J Immunol 1991; 146:3218-3226.
120. Weber JS, Berry J, Manser T, Claflin JL. Position of the rearranged Vκ and its 5' flanking sequences determines the location of somatic mutations in the Jκ locus. J Immunol 1991; 146:3652-3655.
121. Rothenfluth H, Taylor L, Bothwell A, Both G, Steele E. Somatic hypermutation in 5' flanking regions of heavy chain antibody variable regions. Eur J Immunol 1993; 23:2152-2159.
122. Gearhart PJ, Bogenhagen DF. Clusters of point mutations are found exclusively around rearranged antibody variable genes. Proc Natl Acad Sci USA 1983; 80:3439-3443.
123. Goyenechea B, Klix N, Yelamos J et al. Cells strongly expressing Ig-kappa transgenes show clonal recruitment of hypermutation: A role for both MAR and the enhancers. EMBO J 1997; 16:3987-3994.
124. Fulton R, Van Ness B. Kappa immunoglobulin promoters and enhancers display developmentally controlled interactions. Nucleic Acids Res 1993; 21:4941-4947.
125. Tumas-Brundage K, Manser T. The transcriptional promoter regulates hypermutation of the antibody heavy chain locus. J Exp Med 1997; 185:239-250.
126. Rabbits T, Forster A, Hamlyn P, Baer R. Effect of somatic mutation within translocated *c-myc* genes in Burkitt's lymphoma. Nature 1984; 309:592-597.
127. Morse B, South V, Rothberg P, Astrin S. Somatic mutation and transcriptional deregulation of *myc* in endemic Burkitt's lymphoma disease: Heptamer-nonamer recognition mistakes? Mol. Cell Biol 1989; 9:74-82.
128. Mueller J, Janz S, Goedert J, Potter M, Rabkin C. Persistence of immunoglobulin heavy chain/c-myc recombination-positive lymphocyte clones in the blood of human immunodeficiency virus infected homosexual men. Proc Natl Acad Sci USA 1995; 92:6577-6581.
129. Migliazza A, Martinotti S, Chen W et al. Frequent somatic hypermutation of the 5' noncoding region of the BCL6 gene in B-cell lymphoma. Proc Natl Acad Sci USA 1995; 92:12520-12524.
130. Pongubala J, Nagulapalli M, Klemsz S, McKercher S, Maki R, Atchison M. PU.1 recruits a second nuclear factor to a site important for immunoglobulin κ 3' enhancer activation. Mol Cell Biol 1992; 12:368-378.
131. Rudin C, Storb U. Two conserved essential motifs of the murine immunoglobulin λ enhancers bind B-cell-specific factors. Mol Cell Biol 1992; 12:309-320.

132. Eisenbeis C, Singh H, Storb U. PU.1 is a component of a multiprotein complex which binds an essential site in the murine immunoglobulin λ2-4 enhancer. Mol Cell Biol 1993; 13:6452-6461.
133. Taylor L, Carmack C, Huszar D et al. Human immunoglobulin transgenes undergo rearrangement, somatic mutation and class switching in mice that lack endogenous IgM. Internatl Imm 1994; 6:579-591.
134. Szajnert M, Saule S, Bornkamm G, Wajcman H, Lenoir G, Kaplan J. Clustered somatic mutations in and around first exon of non-rearranged c-myc in Burkitt lymphoma with t(8; 22) translocation. Nucleic Acids Res 1987; 15:4553-4565.
135. Ye B, Cattoretti G, Shen Q et al. The Bcl-6 proto-oncogene controls germinal-center formation and Th2-type inflammation. Nature Genetics 1997; 16:161-170.
136. Brenner S, Milstein C. Origin of antibody variation. Nature 1966; 211:242-243.
137. Steele E, Pollard J. Hypothesis: Somatic mutation by gene conversion via the error prone DNA> RNA > DNA information loop. Mol Immunol 1987; 24:667-673.
138. Maizels N. Might gene conversion be the mechanism of somatic hypermutation of mammalian Ig genes? Trends in Genetics 1989; 5:4-8.
139. Manser T. The efficiency of antibody maturation; can the rate of B cell division be limiting? Immunol Today 1990; 11:305-308.
140. Rogerson B, Hackett J, Peters A, Haasch D, Storb U. Mutation pattern of immunoglobulin transgenes is compatible with a model of somatic hypermutation in which targeting of the mutator is linked to the direction of DNA replication. EMBO Journal 1991; 10:4331-4341.
141. Storb U, Peters A, Klotz E, Rogerson B, Hackett J. The mechanism of somatic hypermutation studied with transgenic and transfected target genes. Semin Immunol 1996; 8:131-140.
142. Shen HM, Cheo DL, Friedberg E, Storb U. The inactivation of the XP-C gene does not affect somatic hypermutation or class switch recombination of immunoglobulin genes. Mol Immunol 1997a; 34:527-533.
143. Mellon I, Rajpal D, Koi M, Boland C, Champe G. Transcription-coupled repair deficiency and mutations in human mismatch repair genes. Science 1996; 272:557-560.
144. Storb U, Klotz E, Hackett J, Kage K, Bozek G, Martin TE. A hypermutable insert in an immunoglobulin transgene contains hotspots of somatic mutation and sequences predicting highly stable structures in the RNA transcript. 1998; (submitted).
145. Rogozin I, Kolchanov N. Somatic hypermutagenesis in immunoglobulin genes. II. Influence of neighboring base sequences on mutagenesis. Biochim Biophys Acta 1992; 1171:11-18.
146. Shen H, Peters A, Baron B, Zhu X, Storb U. Mutation of Bcl-6 gene in normal B cells by the process of somatic hypermutation of Ig genes. Science 1998; (in press).

CHAPTER 5

Germinal Center Derived Lymphomas in Humans and Mice

V.K. Tsiagbe, N.M. Ponzio, G.S. Erianne, D.J. Zhang,
G.J. Thorbecke and G. Inghirami

Human Germinal Center-Derived and Related Lymphomas

Introduction

Tumor cells represent the neoplastic counterparts of normal elements which have undergone a series of irreversible genetic alterations. The end result is a cell capable of indefinite proliferation and of escaping many or all of the physiological control mechanisms. In general, the neoplastic cells tend to maintain at least some of the morphological and/or phenotypic features characteristic of the normal cells from which they derive. This also applies to lymphomas. Accordingly, human lymphomas have been classified in different categories and subgroups based on the morphology of the neoplastic cells, their immunophenotype and their capability to follow normal differentiation pathways and to recapitulate some of the lymphoid structure seen in normal lymphoid tissues.

Besides the neoplasms primarily derived from the centrocytes and centroblasts from germinal centers, a large amount of information has been obtained during the last decade demonstrating that several other human neoplasms represent the neoplastic counterparts of unique elements present within the general microenvironment of the germinal center and B cell follicle (Table 5.1). Overall, these neoplasms have been recognized and appropriately characterized on the basis of unique immunophenotypes (Table 5.2) and/or specific genetic abnormalities (Table 5.2). For some lymphomas, however, the lack of specific aberrations or other unique markers

The Biology of Germinal Centers in Lymphoid Tissue, edited by G. Jeanette Thorbecke and Vincent K. Tsiagbe. © 1998 Springer-Verlag and R.G. Landes Company.

Table 5.1. Germinal center characteristics and T cell associations seen in various human B cell lymphomas

Lymphoma Type	Normal B Cell Equivalent	Associated Cells	Apoptosis & Assoc. Proteins	Bcl-6[#] Protein	Somatic[##] Mutation
Burkitt's	Centroblast (Bm3)	$\gamma\delta$[$]	Bcl-2$^{+/-}$, Bcl-X^{hi}, CD95$^-$, c-Myc$^-$	High	~5%
NLP Hodgkin's	Centrocyte (Bm4)	CD4, 20% CD57^{+*}	Bcl-2$^-$, CD95$^+$	High (100%)	Ongoing
Classical Hodgkin's	Not clear	CD4$^+$CD40L$^+$ **	Bcl-2$^+$, CD95$^+$	High (30%)	High, crippling
MALT/Marginal Zone	Memory B (Bm5) ($\mu^+\delta^-$ or γ^+)***	Reactive T Cell Infiltrates & Some Paraprotein	Bcl-2$^+$, CD95$^+$	–	High, ongoing
CLL[@]	Foll. Mantle (Bm2)	Clonal CD4 + CD8 T cell Infiltrates[@]	Bcl-2$^+$, CD95$^-$	–	Very Low-6%
Mantle Cell	Mantle (Bm 1)	Bands of T cells[$$] (possibly autoreactive)	Bcl-2$^+$, Bcl-X^{lo}, CD95$^-$, c-Myc$^+$	–	Very Low
Follicular Center Cell	Centrocyte (Bm4)	Variable amount of T cell Infiltrates*	Bcl-2$^+$, Bcl-X^{lo}, CD95$^{+/-}$, c-Myc$^+$	High	~12% (ongoing?)
Diffuse Large B	Centrocyte/blast ?	"	Bcl-2$^{+/-}$, CD95$^{+/-}$	High	Very High
Multiple Myeloma	Plasma cell			–	~8% (2.7-16.5)%

§ Tsiagbe, VK, unpublished observations.

* A high percentage of CD45RO+CD4+CD57+ T cells (19%) is seen in the nodular lymphocyte predominant (NLP) form of Hodgkin's disease (HD), forming rosette-like rings around the tumor cells.[87,88]

** The classical forms of Hodgkin's disease and follicular lymphomas have far fewer of the CD57+ T cells (< 4%), but frequently contain infiltrates of CD4+CD40L+ T cells.[92]

*** Memory B cells have been shown to localize predominantly in the marginal zone areas.[56,57] Although Waldenstrom's macroglobulinemia is usually closely related to lymphoplasmacytoid lymphoma in the Revised European-American Lymphoma (REAL) classification, rare cases of marginal zone B cell lymphomas are associated with IgM paraproteinemia.[142]

@ Serrano et al[96] have described clonal CD4+CD57+ T cells in the peripheral blood of patients with chronic lymphocytic leukemia (CLL).

§§ Tumor cell nodules are frequently surrounded by bands of T cells in these lymphomas.[91]

& The high grade lymphomas tend to be CD95−, suggesting a tumor-suppressor role for CD95; yet, 2/3 of high grade CD95+ lymphomas lack Bcl-2, suggesting that other regulators besides Bcl-2 may regulate the CD95-mediated apoptosis signal.[135] In contrast, upregulation of CPP32 (caspase-3) appears associated with progression.[136] The apoptotic index is highest among the Burkitt's and immunoblastic diffuse large B cell lymphomas.[137] Bcl-2 is expressed in Reed Sternberg cells from classical forms of HD, but not in the NLP form.[138]

Data from: Falini et al[91] (NLPHD: 100%; classical HD 30%); Carbone et al[139] (AIDS-related diffuse large cell 56%; small noncleaved 100%); Flenghi et al;[140] Miura et al.[141]

Data from: Kanzler et al[143] (crippling, high: Reed-Sternberg cells are GC derived); Braeuninger et al[144] (in NLPHD, somatic mutations are ongoing), see also Marafioti et al[64] and Ohno et al,[65] Bahler et al[145] (MALT salivary gland, ongoing as judged by intraclonal sequence heterogeneity, restricted use of VH and VL gene segments), Klein et al[146] (compare MALT to IgM+IgD− memory B cells which also have mutated V regions); Tamaru et al[45] (Burkitt's 5%, cells are arrested in early centroblast stage because of deregulated *Myc*?); Oscier et al[147] (CLL with trisomy 12: 0.34 +/- 0.86%, but cases with 13q14 abnormality 6.5 +/- 1.7%); Tamaru et al[45] (FCC 11.8% ongoing); Vescio et al[148] (multiple myeloma: 2.7-16.5%, mean 8.2%), see also Baker et al[149] and Bakkus et al[150] (multiple myeloma: high, sometimes preswitched); Wagner et al[151] (Hairy cell: Very Low); Kuppers et al[152] (Diffuse large cell lymphomas have a high load of somatic mutations with evidence of selection for antibody expression, perhaps because they underwent a prolonged stay in the germinal center microenvironment).

Table 5.2. Germinal Center Derived And Related B Cell Lymphomas

Clinical Features[*]	Surface Phenotype				Genotype[**]
	CD19	CD10	CD5	CD23	
Small Lymphocytic[Low]	+	–	+	+	Trisomy 12,13q
Mantle Cell[Low]	+	+/–	+	–	t(11;14) (CyclinD1[Hi])
Marginal Zone[Low] or MALT (saliv., stomach)	+	–	–	–	chromos.# 1,3,7,8[$];trisomy 3
Multiple Myeloma	+	–	+/–	–	t(11;14),t(4;14) (FGF-R3)
Follicular S Cleaved[Low]	+	+	–	+/–	t(14;18) (Bcl-2)
Follicular L Cell[Int]	+	+/–	–	+/–	t(14;18)
Diffuse S Cleaved and L ± Cleaved[Int]	+	+/–[‡]	+/–[‡]	–	t(14;18), t(8;14) (c-Myc) 3q29 (Bcl-6)
Burkitt's[Hi]	+	+	–	–	t(8;14), t(2;8), t(22;8)(c-Myc)

[*] The superscripts reflect the clinical lymphoma grades: Low, Intermediate (Int.) or High (Hi). S = small; L = large.
   The relationship of some of these lymphomas to GCs is discussed in the text.
[**] The genotypic characterization was taken from Harris et al.[153]
[‡] Lymphomas with immunoblastic phenotype are a subset of this group, exhibiting CD10 in ~16% and CD5 in ~8% of cases.
[$] Abnormalities variable, mostly on these chromosomes; some also trisomy 3.[154] These lymphomas usually bear the α4β7 mucosal homing receptor.[155]

(hairy cell leukemia) have so far prevented the identification of a correspon ᵢ normal counterpart. Based on these considerati we will discuss the major lymphomas which may have their normal counterparts within the germinal center and B cell follicles.

Follicular Cell Lymphoma (FCCL)

Many of the B cell neoplasms, including small lymphocytic, plasmacytoid lymphocytic, follicular center cell, immunoblastic plasmacytoid, plasmacytoma, mantle cell and monocytoid B cell, have been recognized on the basis of pure morphologic criteria. This is particularly true for neoplasms derived from normal germinal center cells, which especially in the case of a nodular structure, closely resemble the normal lymphoid cell type from which they are derived.[1,2] Rappaport has clearly defined the criteria to distinguish FCCL from normal follicular hyperplasia.[3] However, the precise relationship between these entities and their normal cellular counterparts was recognized only in the early 1970s.[4-6] The recognition that the cells of FCCL morphologically resemble the normal elements of germinal centers allowed the identification not only of the classical nodular forms of these lymphomas, but also the diffuse types.

Cytologically, the nodules can be composed of centrocytes ("cleaved" cell) and centroblasts ("noncleaved"). According to their composition they are classified as small/centrocytic or large/centroblastic lymphomas. In these neoplastic nodules the neoplastic cells represent the majority of the total elements; however, normal follicular dendritic cells and intrafollicular T lymphocytes are commonly seen. In some instances, a considerable number of plasma cells and/or plasmacytoid elements are identified which may represent neoplastic cells with plasmacytoid differentiation or normal bystander plasma cells. Notably, mitoses are rare and phagocytic histiocytes are absent. Finally, the cells are arranged uniformly and the normal polarity of germinal centers is lost. In the mixed and diffuse forms the nodules are partially preserved or absent, respectively. The neoplastic cells in the diffuse areas are similar to those described in the nodules and these neoplasms are classified as small cell (SC <25% centroblasts), mixed cell (MC) and large cell types (LC >50% centroblasts). The division of FCCL in SC, MC and LC has important prognostic implications. Patients with small cleaved lymphomas, despite the fact that they are often stage III or IV at presentation, generally have a favorable prognosis and long term survival (7-9 years).[7] This may be due to the fact that the neoplastic cells have relatively low mitotic rates.

FCCL are derived from mature peripheral B cells which have undergone Ig H and L chain gene rearrangements. Using Southern blot analysis virtually all FCCL show gene rearrangement products (100% VH, 100% Vκ and 30% Vλ genes). Recently, polymerase chain reaction (PCR) analysis has been used to prove the clonal origin of these lymphomas.[8] When follicular lymphomas are analyzed for the expression of sIg, more than 50% of the cases show sIgM, with only a minority also expressing sIgD.[9] FCCL express sIgG more often than other low grade B cell lymphomas, such as the CD5+ chronic lymphocytic leukemia (CLL) or mantle cell lymphomas, which are usually IgM+. This frequency of Ig heavy chain class switching is in agreement with their germinal center origin. In virtually all FCC lymphomas, κ or λ light chains are expressed, but in a minority of cases no detectable sIg is seen. The neoplastic B cells express a large variety of B cell associated antigens, including CD19, CD20, CD22,

CD21 (Table 5.2). However, centrocytic/centroblastic lymphomas do not express CD5.[10,11] This is an important criterion for the classification of FCCL and may have implications for the identification of their precursor cells. In ~60% of cases, CD10 (CALLA) can be detected on the neoplastic cells.[10-12] The expression of these surface markers is particularly important in the classification of diffuse and/or large cell lymphomas and allows the identification of their germinal center origin. The neoplastic cells do not express any T cell markers, but infiltrating T cells are often seen (Table 5.1). Typical follicular dendritic cell (FDC) surface markers, such as CR1,[13] are also seen within the neoplastic nodules in ~30-40% of the cases,[14,15] demonstrating the presence of FDCs (which are probably not neoplastic themselves?).

Multiple genetic abnormalities are generally necessary for the development of totally transformed neoplastic cells. The systematic utilization of cytogenetic and/or molecular analyses have allowed the discovery that unique neoplastic processes are associated with specific genetic abnormalities. In the case of FCCL, the most common genetic abnormality is a translocation involving the long arms of chromosomes 14 and 18; t(14;18)(q32;q21).[16-18] This translocation involves the loci of the Ig H chain and the *bcl-2* genes. In the case of the Ig gene, the breakpoint usually occurs in the J region. The breakpoint in chromosome 18 is more variable and different breakpoints have been identified. The most common one (mbr) is clustered within a 150 bp sequence in the 3' untranslated region of the third exon of the gene.[19,20] The translocation of *bcl-2* juxtaposed to the Ig ge   ocus results in deregulation of the *bcl-2* transcript and overexpression of Bcl-2. This is most likely due to the fact that the *bcl-2* gene is now under control of Ig H chain regulatory elements. *bcl-2* is a member of a group of genes involved in cell survival and has been shown to have potent anti-apoptotic properties. The function of Bcl-2 has been partially determined by transfection experiments.[18,21,22] Introduction of a deregulated form of the *bcl-2* gene in interleukin-dependent or myeloid cell lines results in protection of these cells from apoptosis, upon withdrawal of the interleukins.[21,22] It has been postulated that overexpression of Bcl-2 in FCCL may give the neoplastic cells a survival advantage rather than an increase in proliferation. The immunophenotypic demonstration of Bcl-2 has assumed an important role in the classification of lymphomas. By immunohistochemistry Bcl-2 can be identified in resting B and T cells, but, as described above, not in normal germinal centers.[23] On the other hand, all SC cleaved nodular FCCL and a considerable fraction of LC FCC (50-70%) show overexpression of Bcl-2.[24]

In view of the fact that Bcl-2 is incapable of inducing cellular transformation, several authors have speculated that FCCL may carry an additional, and so far unknown, genetic lesion. Interestingly, 25% of FCCs will undergo transformation (diffuse large cell lymphoma or acute leukemia) sometime during the course of the disease. The transformation of these low to high-grade lymphomas has been associated with the additional occurrence of a loss of p53 suppressor gene function.[25-28] Several groups have demonstrated that the loss of p53 is responsible for the morphologic and biologic transformation of more than 80% FCCL. Transformation can also occur when FCCL acquire the t(8;14) translocation, bringing the *c-myc* gene in juxtaposition to IgH.[29] Interestingly, these neoplasms have morphologic features similar to Burkitt's lymphomas (BLs).

BLs or "small, noncleaved cell lymphoma" were first identified by Lukes and Collins,[30] who suggested that these lymphomas may represent the neoplastic counterpart of normal germinal center cells. This hypothesis is further supported by immunophenotypic similarities between classical FCCL, Burkitt's and germinal center cells (CD10$^+$, CD77$^+$).[31] Whether BLs represent germinal center derived neoplasms is still a controversial subject, requiring further investigation.[32,33] It should be noted that the t(8;14) translocations seen in endemic and sporadic BLs involve different IgH chain regions and appear to occur in early and late stages of B cell differentiation, respectively.[34-37] Thus, it is possible to speculate that germinal center cells, while proliferating, may attempt Ig gene switching and undergo translocation, producing either the t(8;14) (*c-myc*) or the t(14;18) (*bcl-2*) translocation. This model tends to support the hypothesis that at least sporadic BLs may be derived from germinal center cells. The identification of neoplasms carrying both translocations tends to further support this proposal.[29] A subset of classical FCC lymphomas carries a t(8;14) (q24;q32) translocation, involving the loci of *c-myc* and Ig H chain instead of the classical t(14;18).[38] Molecular analysis of these cases tends to suggest that this translocation involves regions different from those of the classical t(8;14) seen in BLs, and it is possible that different abnormalities (breakpoints) involving *c-myc* and Ig genes may result in the generation of morphologically different entities, resembling on one hand the classical FCCL, and on the other, BLs.

Small Noncleaved Cell Lymphoma

Small noncleaved cell lymphoma includes Burkitt's lymphoma (BL) and the undifferentiated non-Burkitt type. The precise origin of these neoplasms is still unclear. As discussed above, a considerable number of findings tends to indicate their germinal center origin (Tables 5.1-5.3). BL contain relatively few infiltrating T cells, including some γδ T cells (Tsiagbe VK, unpublished observation). BL cells make a variety of cytokines, of which TNF-α and LT have been associated with an enhanced rate of tumor progression[39] (Table 5.3). Morphologically, BL presents a monotonous proliferation of cells of intermediate size, round to oval nuclei with coarse chromatin and multiple nucleoli. The neoplastic cells have a very high proliferative activity and undergo a high degree of spontaneous cell-death through apoptosis (Table 5.1). Morphologically, these activities are characterized by a high mitotic index and by the "starry sky" appearance. Two major entities, sporadic and endemic BL, have been described. These have quite unique pathogenesis and molecular genetic lesions. The endemic ones virtually always carry EBV (95%) and t(8;14, 2;14 and 8;22) translocations which involve the JH and proximal *c-myc* regions. On the other hand, the sporadic ones generally do not carry EBV and the translocation involves the switch region of γ. The juxtaposition of the Ig enhancer to *c-myc* leads to the deregulation of *c-myc* which acts as a potent oncogene. The immunophenotype of BL cells is characterized by the expression of surface κ- or λ-chains,[13] IgH chain (μ, but little δ),[40-42] CD10, CD77,[43] CD38, CD19 and CD20, but CD21 is often undetectable.[44] The occurrence of a definite, but relatively l   rate of somatic mutations in the IgVH regions s gests an early centroblast origin of BL cells.[45] Furthermore, several activation markers are expressed by the neoplastic cells in agreement with their high rate of cell proliferation. Notably, the lack of LFA-1 and LFA-3 may play an important role in tumor escape from the host immunosurveillance.[46,47] Clinically, the endemic forms of BL are well characterized. On the other hand, the differences between the classic forms of BL and non-Burkitt small noncleaved cell lymphomas among the sporadic ones (HIV$^+$ and HIV$^-$) are ill defined. In general the sporadic BLs and non-BLs tend to occur in patients 6.5-35 years of age for BL and at 24.5-67 years of age in the non-BL. The two groups appear to have similar clinical presentations, including stage, symptoms, bone marrow involvement[48,49] and prognosis.[50]

Table 5.3. Cytokine production by B cell lymphomas

Lymphoma Type	Cytokines Produced	Reference
Hodgkin's Reed-Sternberg Cells (1/1)	IL-1, IL-5, IL-6, IL-9, M-CSF, TGF-β	Hsu et al[156]
(3/3)	IL-9	Gruss et al[60]*
(6/13)	IL-9	Merz et al[157]
Burkitt's $		
(+ anti-Ig 3/3)	TNF-α	Goldfeld et al[158]
(6/10)	LT	Gibbons et al[39]@
(5/8)	IL-13	Fior et al[159]
(5/8)	IL-13	Emilie et al[160]
(10/11)	IL-6	Emilie et al[161]&
Am.(10/16), Afric. (1/5)	IL-7	Benjamin et al[162]
(EBV+ only, 9/9)	IL-10	Benjamin et al[163]#
(3/3)	GM-CSF	Corcione et al[164]
Follicular Center (5/8)	IL-13	Fior et al[159]
(5/8)	IL-13	Emilie et al[160]
(3/3)	GM-CSF	Corcione et al[164]
Diff. Large Cell (3/7)	IL 13	Fior et al[159]$
(3/7)	IL-13	Emilie et al[160]
(7/7)	LT	Gibbons et al[39]@
(6/7)	IL-6	Emilie et al[161]&
(2/6)	IL-9	Merz et al[157]
B-CLL (9/12)	IL-1	Pistoia et al[165]
(3/3)	IL-13	Fior et al[159]
(3/3)	IL-13	Emilie et al[160]
–SAC (7/14); +SAC (4/4)	GM-CSF	Zupo et al[166]
+SAC (15/15)**	G-CSF	Corcione et al[167]

Burkitt's lymphoma, follicular cell lymphoma and diffuse large cell lymphoma are thought to have GC cells as their normal counterpart in a significant percentage of cases; B cell chronic lymphocytic leukemia is thought to have the CD5+ follicular mantle B cell as its normal counterpart.[168]

* CD30+ cells isolated from the patients exhibited mRNA for IL-9. This is an autocrine factor for Reed-Sternberg cells, as they also proliferate in response to IL-9.[60]

\# In view of the high degree of homology between the viral product BCRF1 and human IL-10, the identity of the product of the lymphoma cells was not firmly established.[163]

$ Am. = American; Afric. = African.

& Burkitt's lymphoma cells make much less IL-6 than diffuse large cell lymphoma.[161]

@ Autocrine factor; both the TNF-R and LT were expressed primarily by Burkitt's lymphomas which developed a lymphoblastoid phenotype.[39]

** Unstimulated (-SAC) or stimulated with *S. aureus* C (+SAC).

In general, the BL tend to have higher extranodal (gastrointestinal tract) presentation compared to the non-BL, which often present with generalized lymphadenopathy.

Mantle Cell Lymphoma (MCL)

Mantle cell lymphoma is a new group of neoplasms which includes all those lymphomas previously defined as "intermediate differentiated lymphocytic", "centrocytic of the Kiel classification" and "mantle zone lymphomas". Morphologically it is characterized by the presence of intermediate-small lymphocytes, with irregular nuclear contours and rather open chromatin. Essential criteria are the presence of a high mitotic index throughout the tumor. The neoplastic cells may grow as a diffuse process and/or in nodular fashion. Rarely, these neoplasms transform into a more aggressive variant.[12,51] The immunophenotype is quite unique and extremely useful in the characterization and identification of this process. The neoplastic cells express intermediate-high levels of surface Ig (IgM) and, in addition to B cell specific markers, they express in most cases CD5, CD43 and CD6, while generally lacking CD10 and CD23. Interestingly, the putative normal counterpart, the lymphocytes of the mantle zone, express IgM, IgD and CD5 and, like MCL, lack somatic hypermutations. The distinct feature of these neoplasms is the deregulation and consequent overexpression of cyclin D1. This is generally achieved through the juxtaposition of IgH chain enhancer (Eµ) sequences to those of the *bcl-1/PRAD-1* gene as a consequence of the t(11;14) translocation.[52-54] Cyclin D1 overexpression has been associated with .vth deregulation and tumor formation. The clin: presentation is similar to that of other small B cell neoplasms: They occur at a median age of 60 years, predominantly in males, without symptoms, presenting with generalized lymphadenopathy and mostly (~80%) in stage III or IV. Similar to that for patients with other low-grade lymphomas, the survival curve of these patients demonstrates a continuous slope, without plateau, and a survival range of 36 to 56 months.

Diffuse Large B Cell Lymphoma (DLCL) and Marginal Zone B Cell Lymphoma

Diffuse large cell lymphoma comprises two major categories of neoplasms: large B cell tumors derived from follicular center cells and immunoblastic lymphomas of B or T cell origin. Morphologically, the B cell neoplasms are quite heterogeneous. Based on the cytological features, multiple subtypes have been described (e.g., large noncleaved, large cleaved, etc.). Immunophenotypically, these neoplasms bear B cell associated antigens (CD19, CD20 and CD22) in virtually all the cases (Table 5.2). However, a considerable fraction of them do not express detectable L and/or H Ig chains. Finally, the expression of CD10 (7-40%) and CD5 (0-14%) indicates that approximately a third derive from germinal center cells and only a minority represent the transformation of CD5$^+$ B cells. It should also be considered that a fraction of the neoplasms arising in extranodal sites (mucosal associated organs) represent the transformation of low grade MALT lymphomas or maltomas.[55] These neoplasms, and the related monocytoid lymphomas, are most likely derived from the marginal zone B cells. Memory B cells, which most resemble the cell type in these lymphomas (Table 5.1), have been shown to localize predominantly in the marginal zone areas.[56,57] The current hypothesis is that a chronic antigenic stimulation (*Helicobacter pylori* in the cases of gastrointestinal neoplasm) leads directly or indirectly to chronic stimulation of T and B cells, allowing the B cell outgrowth and possibly the accumulation of genetic errors, which may ultimately be responsible for the uncontrolled growth.[58] The acquisition of genetic defects such as p53 mutations then allows the full transformation of these neoplasms from low-grade into high-grade.

Several genetic alterations have been described among DLCL (Table 5.2). The most common ones are the t(14;18) and t(8;14) translocations; the former most commonly in nodal types and the latter in non-nodal tumors. The reported overall frequencies of t(14;18) and t(8;14) are 12%-40% and 35%, respectively. Rare neoplasms carrying both alterations have also been described. Finally, several groups have recently demonstrated that a considerable proportion of DLCL carries "promiscuous" translocations involving 3q27. This is the locus of *bcl-6*, which can be found juxtaposed to a large number of alternative genes.[59] In fact, only in approximately 50% of cases is the Ig gene locus translocated to *bcl-6*. In those cases, Ig gene switch regions are mostly involved, indicating that the translocation may occur late in B cell development and differentiation.

DLCLs represent a distinct clinical entity with a wide patient age range, often extranodal sites, generally localized initial presentation, but often with rapid progression. The overall survival of these patients has considerably improved, since the usage of high dose multi-drug therapy and complete remissions are now achieved in approximately 80% of the patients.

Hodgkin's Disease (HD)

Multiple approaches have recently been used to study the nature, lineage, and clonality of Reed-Sternberg (RS) cells and their variants, the malignant cells of HD. Many hypotheses have been generated from these studies. The most popular current hypothesis is that RS cells represent activated B cells capable of elaborating (and/or eliciting!) a large number of cytokines (Table 5.3). These are responsible for activating bystander lymphocytes and cellular connective tissue elements to form the pathologic lesion we commonly recognize as HD. In addition, IL-9 has been identified as an autocrine factor for RS cells.[60] The preferential membrane and/or cytoplasmic expression of B cell associated or restricted antigens in RS cells supports the hypothesis that RS cells in different histologic types of HD may be preferentially derived from B cells. Using single cell PCR, several groups have recently demonstrated that RS cells often carry IgH chain gene rearrangements[61,62] and many somatic mutations (see Table 5.1). However, the same clonal products are not always identified when different RS cells derived from the same patient are investigated.[63] The most current opinion is that RS cells are clonal in both the "nodular lymphocyte predominant" form of HD (NLPHD) and classical HD, but that due to the ongoing somatic mutations in the NLPHD form, intraclonal diversity arises.[64,65] Interestingly, Bcl-6 expression has been demonstrated not only in RS cells of NLPHD, but also of some classical HD (Table 5.1). These findings tend to support the hypothesis that RS may derive from B cells which have at one time passed through a germinal center. Due to the difficulties in obtaining the RS cells, classical cytogenetic approaches have rarely succeeded and no consistent pathognomonic chromosomal aberrations or specific defects could be identified among HD cases.[66] By using PCR, *bcl-2* rearrangements have been found in some HD cases[67] and the transformation of classical HD into large cell lymphoma, carrying *bcl-2*-Ig translocation, has also been documented.[68]

Unique Properties of Certain Germinal Center-Derived Lymphomas

CD30

Germinal center formation and germinal center cell survival are highly regulated phenomena. These processes require the coordinated expression and activities of a large number of molecules (chapter 1). Among these players, several antigens belonging to the TNF superfamily, such TNF, LTα, FAS, CD40L, CD40 etc. have been demonstrated to be present[69] and to play a crucial role in germinal center physiology (see chapter 1). However, it is possible that other members may have a role which has not yet been identified. CD30 has recently been demonstrated to play an important role in thymic negative selection,[70] but there is so far no evidence that it is involved in germinal center formation or function. On the other hand, using highly sensitive immunohistochemical detection systems, we and others have recently demonstrated that CD30[+] cells can be demonstrated within the germinal center, in addition to interfollicular areas. The precise function and nature of these cells is still unknown. However, it has become clear that a subset of human and mouse neoplasms derived from germinal centers express surface CD30. This is particularly the case for the RS cells of Hodgkin's disease and the germinal center-derived lymphomas of SJL mice. Although a large percentage of high grade B cell lymphomas express CD30L, the presence of both CD30 and CD30L in the same tumor, suggesting the possible interactions of cells within the same lymphoma, has not yet been clearly established.[71]

Reexpression of Proteins Concerned with Ig Rearrangement

B and T lymphocytes undergo a series of complex and highly controlled events leading to the generation of functional antigen receptors. In order to generate their surface IgM and T cell receptors these cells have to express a large number of gene products necessary for the antigen receptor gene rearrangement, in a timely and coordinate fashion during differentiation. It has been postulated that the expression of genes involved in B cell ontogeny, such as RAG-1, RAG-2 and terminal deoxyribonucleotidyl transferase (TdT) was exclusively restricted to the early phase of B cell development. On the other hand, new evidence has emerged on the expression of these genes in cells which have already undergone gene rearrangement. Several authors have demonstrated that these products are not only present in neoplastic cases derived from early precursors, such as lymphoblastic leukemia, but also in Burkitt's lymphoma,[72] non-Hodgkin's lymphoma and leukemia which represent the neoplastic counterparts of mature, fully differentiated lymphoid cells.[73,74] The aberrant expression of these genes could of course be a consequence of the neoplastic transformation. However, recent findings indicate that mature B cells indeed have the capacity to reexpress several genes which are involved in early stages of B cell development. In particular, RAG-1 and RAG-2 mRNA transcripts and corresponding proteins can be found in activated B cells and in normal germinal center cells.[75-77] The functional significance of this reexpression with respect to receptor editing by such B cells is discussed in chapter 1. Based on these new findings, one can speculate that some B cell neoplasms may represent the neoplastic counterpart of these cells, frozen in this particular and unique stage of differentiation. If this hypothesis is correct, these neoplasms may be able to undergo further Ig gene rearrangements. Preliminary observations suggest that Reed-Sternberg cells of Hodgkin's disease are a plausible candidate for such neoplasms, as they express VpreB and λ-like chains (Prolla G, Ponzoni M, Mearson A, Chiarle R, Inghirami G, unpublished observations).

Bcl-6

Recently, several groups have demonstrated that *bcl-6* (located at 3q27) is often translocated in diffuse LC and in 5-10% of FCC lymphomas[78-80] leading to a high expression of Bcl-6 protein in the nuclei of such tumor cells (see Tables 5.1 and 5.2). Based on the cDNA sequence, Bcl-6 may function as a DNA-binding transcription factor that may be involved in B cell proliferation and/or differentiation.[81] Using polyclonal antibodies specific for internal peptides of Bcl-6, it has recently been demonstrated that high levels of Bcl-6 protein can be detected in B cells within germinal centers.[81,82] Both centrocyte d centroblasts appear to express the Bcl-6 protein a , interestingly, CD4+ T cells in germinal centers also show nuclear staining, suggesting that Bcl-6 may play an important role in germinal center physiology and in the regulation of memory B cell differentiation (see chapter 1).

T Cells Associated with Germinal Center-Derived Lymphomas

In the germinal center-derived lymphomas of mice (see below), T cells from the host, responding to a retroviral superantigen (vSAg) expressed by the lymphoma cells, provide cytokines which promote lymphoma growth. Further investigation is needed to determine whether this phenomenon of reverse immune surveillance in host-lymphoma interaction, on the basis of the expression or presentation of an antigen or SAg by lymphoma cells, is of relevance to any of the human germinal center-derived lymphomas. Although it is well known that human B cell lymphomas often contain T cell infiltrates, the specificity and functional significance of these T cells is unknown.[83-86]

CD45RO$^+$CD4$^+$CD57$^+$ T cells, which are of memory phenotype and NK-like, are under normal conditions typically found in the light zone of GCs. A high percentage of such cells (19%) is seen in the nodular lymphocyte predominant form of HD, forming rosette-like rings around the tumor cells.[87,88] These cells lack expression of CD26[89] and CD40L[90,91] and express Bcl-6 in their nuclei.[91] The classical forms of Hodgkin's disease and follicular lymphomas have far fewer of the CD57$^+$ T cells (< 4%), but frequently contain infiltrates of CD4$^+$CD40L$^+$ T cells.[92]

As mentioned in the discussion of marginal zone B cell lymphomas, responsiveness of T and/or B cells to *Helicobacter pylori* is thought to play a role in the pathogenesis of stomach (MALT) lymphomas, as eradication of the *H. pylori* is associated with regression of low grade (but not of high grade) MALT lymphomas.[93] In addition, heat-killed *H. pylori* bacteria stimulate proliferation of the infiltrating T cells and, in the presence of those T cells, also of the tumor cells in vitro.[93a+b] In mantle cell lymphomas, T cell infiltrates are frequently present as bands surrounding tumor cell nodules.[94] In some patients with these lymphomas, the T cells have a memory phenotype, and are stimulated by autologous tumor cells to induce polyclonal B cell differentiation.[95]

With respect to lymphomas not derived from germinal centers, elevated CD4$^+$CD57$^+$ T cells in PBL correlated with advanced disease, suggesting a pathogenic role of these T cells in chronic B lymphocytic leukemia.[96] In contrast, the CD8$^+$ T cells are thought to be associated with resistance to growth of these lymphomas, as after therapy with anti-CD52 (Campath-1H), long term tumor free remissions are associated with detection of monoclonal CD8$^+$CD52$^-$ T cells.[97] In addition, Serrano et al[96] have described a restricted Vβ repertoire for the CD4$^+$ T cells associated with these lymphomas. However, similar studies on the T cells infiltrating germinal center-derived lymphomas have not been reported to our knowledge.

Germinal Center-Derived Lymphomas of SJL Mice

Characteristics of SJL Lymphomas

Primary lymphomas in SJL mice represent the only well characterized model for germinal center-derived lymphomas in mice.[98-101] These lymphomas, originally called reticulum cell sarcomas or RCS, affect >95% of SJL mice by the age of 13 months. RCS lymphomas develop to the same frequency in aging male and female SJL mice and to the same extent in germfree and conventional SJL mice,[102] suggesting the absence of any gender-specific or strong environmental influences on RCS development. RCS lymphomas are derived from B cells which exhibit switched Ig isotype gene patterns.[103] A large percentage of SJL mice of 6 months of age or older exhibit single or multiple paraproteins in their sera with an isotype distribution of $\gamma 1 \geq \gamma 2a >> \gamma 3 >> \gamma 2b$ (no μ, δ, or ϵ).[104,105] The most studied transplantable and/or tissue culture RCS-lines developed from primary SJL lymphomas, cRCS-X and cRCS-2, do not exhibit surface Ig, however, because of errors made during switching affecting the H chain transcription.[103,105] Nevertheless, both of these lines secrete κ chains, and primary SJL lymphomas do occasionally exhibit sIg. In one documented example, RCS-4, which exhibited surface IgG1 as a primary lymphoma, the cells were found to secrete IgE after several weeks in culture in the presence of γ-irradiated normal syngeneic lymph node cells. This finding suggests that cRCS-4 had continued to undergo isotype switching in vitro (Tsiagbe VK, Ovary Z, Thorbecke GJ, unpublished observations). RCS cells exhibit B220, CD40 and I-A^s, further att ing to their B cell origin. In addition, in view of presence of CD30 on Hodgkin's lymphoma cells and most diffuse large cell lymphomas in humans, SJL lymphomas were examined for and found to express this antigen. Since ligation of CD30 can be involved in the induction of apoptosis, {as suggested by studies on CD30 transgenic mice (Chiarle R, Podda A, Prolla G, Podack ER, Inghirami G, Thorbecke GJ, unpublished observations) and on CD30$^{-/-}$ mice[70]}, the effect of anti-CD30 on lymphoma growth in vivo and in vitro was examined. Anti-CD30 did not affect RCS cell proliferation in vitro, neither in soluble nor in immobilized form. However, injection of anti-CD30 a few hours before lymphoma cells caused a highly significant inhibition (> 50 %) of lymphoma growth in vivo (Simmons WJ, Tsiagbe VK, Charyton D, Podack ER, Thorbecke GJ, unpubl. observ.). The mechanism of this effect is still under investigation.

Peyer's patches and lymph nodes from aging SJL mice which have not yet developed obvious lymphomas occasionally show abnormally shaped extensions of PNA binding cell areas characteristic of germinal centers. This is seen in ~10% of SJL mice of 6-11 months of age and has been interpreted as a prelymphomatous change. These observations, as well as the PNA binding property of overtly abnormal germinal centers in aging SJL mice with beginning lymphomas, seen at greater frequency (75%) in 6-11 months old *bcl*-2 transgenic SJL mice,[101] reinforce the original conclusion of Siegler and Rich[98] that RCS lymphomas are germinal center-derived (Fig. 5.1).

T Cell Dependence of SJL Lymphoma Growth

The growth of transplantable lymphomas and/or incidence of primary lymphomas is diminished in immunocompromised SJL hosts,[106-108] in thymectomized, γ-irradiated and bone marrow reconstituted,[109] in athymic,[108] and in anti-CD4-treated normal SJL mice.[110,111] Transplantable RCS growth in γ-irradiated syngeneic hosts is reconstituted by addition of normal lymph node cells.[107] Development of in vitro lines from these lymphomas requires the addition of γ-irradiated syngeneic lymph node cells at regular intervals.[112] This dependence of RCS growth on normal lymphoid cells of the host has been dubbed "reversed immunological surveillance".[113] Syngeneic CD4$^+$ T cells respond to irradiated RCS cells with strong proliferation and copious production of various cytokines, including IL-2, IFN-γ, IL-4 and IL-5 (reviewed in Ponzio et al[113]). Indeed, the initial in vitro growth of these lymphomas appears dependent on cytokines produced by CD4 T cells.[114] When the responses of these lymphomas to cytokines were examined, it was found that they proliferate in response to IL-5, IL-4, and human LMW-BCGF, in synergy with other cytokines such as IL-1 and IFN-γ. It was also found that IL-4 injections into SJL mice accelerate the development of primary lymphomas. IL-4 not only increases the tumor incidence but also renders SJL mice susceptible to lymphoma development at an earlier age.[115] In contrast to the effects of IL-4 on EAE development,[116] this effect of IL-4 on RCS development in vivo is not counteracted by IFN-γ.

Cytokines produced by the tissue culture lines prepared from SJL lymphoma cells have recently been more extensively characterized by multiprobe ribonuclease protection. Most, if not all of the lines contain mRNA for IL-2, IL-6, IL-15, IFN-γ, LTα, LTβ, TNF-α,

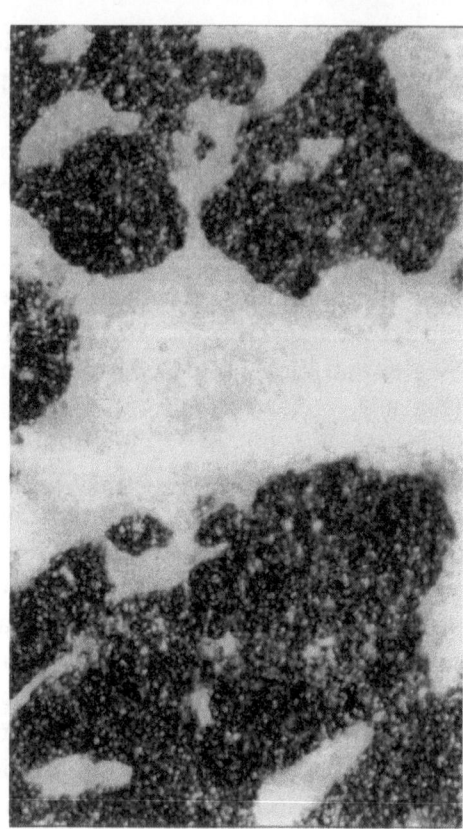

Fig. 5.1. Prelymphomatous, irregularly growing PNA⁺ blast cell areas in Peyer's patch from 11-month-old SJL mouse without overt lymphoma. Staining: PNA-peroxidase and methylgreen. x80.

TGF-β1 and MIF (Tsiagbe VK, Huang C, Zhang DJ, Thorbecke GJ, unpublished observations). In addition, several lines have mRNA for IL-4, IL-5, ⁞ 9, IL-10 and/or IL-13. IL-1 and GM-CSF have not been studied. In view of the results for cytokine production by human B cell lymphomas, summarized in Table 5.3, it is not surprising to find such an array of cytokines produced by RCS cells. The prominence of LT and TNF is particularly noteworthy, since among the human lymphomas, only those most clearly resembling centroblasts appear to produce those cytokines. In view of the absence of germinal centers in LTα or β⁻/⁻ and in TNF-α⁻/⁻ mice, and the suggestion from the work of Dr. Chaplin and coworkers[117] that B cells from LT⁺/± mice induce FDC formation in LT⁻/⁻ recipients, it is of interest to determine whether germinal center derived lymphoma cells are also capable of interacting with FDCs and whether their growth is influenced by this interaction. In this respect, it is noteworthy that soluble LTβ-receptor linked to human Ig partially inhibits the growth of transplantable RCS in SJL mice (Simmons WJ, Tsiagbe VK, Browning JL, Thorbecke GJ, unpublished observations). Another cytokine of great interest is IL-9, because of its possible role as an autocrine factor for Reed-Sternberg cells and diffuse large cell lymphomas (see Table 5.3).

Superantigen Presentation by SJL Lymphomas

T cell clones[118] and hybridomas[119] prepared from responding syngeneic CD4$^+$ T cell populations all bear Vβ16 in their TCR. This discovery of a superantigen-like response led to the cloning and sequencing of an endogenous mouse mammary tumor virus long terminal repeat (MMTV-LTR) gene in SJL lymphomas,[120] later identified by its strain distribution as *Mtv29*.[115,121] Extraordinarily high expression of a 1.8 kb mRNA for this mouse mammary tumor virus long terminal repeat (MMTV-LTR) was found in both primary lymphomas and in in vitro RCS lines but not in an SJL B cell lymphoma, NJ101, that does not stimulate syngeneic T cells.[120] It was demonstrated that the product of this 1.8 kb mRNA serves as a superantigen (vSAg-29) that mediates the vigorous stimulation of SJL host CD4$^+$Vβ16$^+$ T cells by RCS. It is therefore likely that the normal process of RCS lymphomagenesis involves expression of the vSAg-29 by the tumor cells and a response to it by Vβ16$^+$ T cells. Experiments are underway to determine whether expression of *Mtv29* is activated in normal germinal centers of SJL mice. Activation of the transcription of the 1.8 kb mRNA for the vSAg-29 could potentially generate target cells which expand on interaction with Vβ16$^+$ T cells in germinal centers. Characterization of the 1.8 kb mRNA for *Mtv29* in RCS has also shown that the vSAg encoded by this MMTV-LTR does not initiate in the 5' LTR as is typical of many other MMTVs.[122,123] Its initiation is within the *env* region[124] under control of a promoter/enhancer in the *env* region, called META, similar to that originally described for the MMTV product transcribed in EL-4 cells.[125]

Lymphomas of C57L/J Mice

Primary Lymphoma Incidence and Age of Onset

In view of the role of *Mtv29* in the development of lymphoma in SJL mice,[120] it was of interest to examine the lymphomas occurring in the other *Mtv29* bearing strains: C57L/J and Ma/MyJ.[121] Reports in the older literature that C57L/J mice spontaneously develop lymphomas with a frequency from 25%[126] to 55%[127] prompted us to monitor tumor development in aging mice of this strain. C57L/J mice share several other immunological characteristics with SJL/J mice, which suggest that the lymphomas that arise in these two strains may also have a similar etiology. Both strains fail to express MHC class II I-E determinants due to a lack of the I-E α chain genes.[128] Likewise, both SJL and C57L mice have a genetic deletion of TCR Vβ genes which results in a failure to express about 50% of the known murine Vβ chain specificities in their TCR repertoire.[129]

The development of lymphomas in 30 C57L mice was therefore monitored. Seven of these mice (23%) developed progressive enlargement of the spleen and mesenteric lymph nodes, which was first noted at a mean of 21 months of age, some 8 months later than the appearance of lymphomas seen in the majority of SJL mice. Unlike the primary lymphomas that arise in SJL mice, however, no enlargement of the peripheral lymph nodes was noted in any of the C57L mice that developed primary lymphomas.

Phenotypic Characteristics

With respect to cell surface phenotype, the C57L lymphomas are also similar to those that arise in SJL mice. Phenotypic analysis of an in vitro cell line, cNJ123.5, which was derived from a primary C57L lymphoma, revealed expression of MHC class II I-A (but not I-E) determinants, as well as B cell markers, including B220 (CD45R), CD45RA (14.8), CD24 (J11d), and surface IgG. The cNJ123.5 cells also express adhesion molecules [CD11a(LFA-1), CD44 (Pgp-1), and CD54 (ICAM-1)], as well as the co-stimulatory molecules B7-1 (CD80), B7-2 (CD86), and CD24, but are negative for T cell markers CD5, Thy-1.2, CD4 and C . Like SJL lymphomas, this C57L lymphoma also presses CD30. Based on these results, it is clear that the lymphomas which arise spontaneously in aging C57L mice, like the SJL lymphomas, are of B cell lineage, most likely of germinal center origin, particularly as the histology of C57L lymphomas closely resembles that of the SJL lymphomas.

As described above, primary SJL lymphomas are accompanied or even preceded by the appearance of paraproteins in the sera of the mice.[104,130] In a preliminary search for paraproteins in the sera from aging C57L mice, abnormal increases in IgA were found in a few of the sera so far examined, suggesting that paraproteins of that isotype were present.

vSAg Expression and Ability to Stimulate Proliferation of Syngeneic T Cells

C57L B lymphoma cells stimulate vigorous proliferation of lymphoid cells obtained from lymph nodes or spleen of young syngeneic mice. A representative experiment (Fig. 5.2) demonstrates the high degree of proliferation induced by γ-irradiated (γ-) cells obtained from the in vivo transplantable or in vitro NJ123.5 cell line in comparison to the response to an alloantigenic stimulus. This lymphoma-induced proliferative response can be totally abrogated by inclusion of either anti-MHC class II (I-A^b)-specific mAb or anti-CD4 mAb in the co-culture of C57L lymphoid cells and γ-NJ123.5 cells. Thus, just as is seen in the SJL lymphoma model (see above), the C57L B lymphoma cells stimulate proliferation of CD4$^+$ T helper cells via expression of "antigen" in the context of MHC class II. In addition, like the RCS cells from SJL mice, this lymphoma exhibits mRNA for several cytokines, including IL-2, IL-4, IL-6, IL-9, IL-10, IL-13, IL-15, IFN-γ, LT-α, LT-β, TNF-α and TGF-β1 (Tsiagbe VK, Huang C, Ponzio NM, Zhang DJ, Thorbecke GJ, unpublished observations).

To further characterize the nature of this stimulatory ability of C57L lymphoma cells, they were co-cultured with T cell hybrids that express different Vβ specificities in their TCR and, after 24 hours, the supernatants were tested for the presence of IL-2. The data in Table 5.4 demonstrate that cNJ123.5 cells preferentially stimulate IL-2 production in T cell hybrids which express TCR Vβ16. The NJ117 B lymphoma cells, used as a positive control in these experiments, are of SJL origin, and express the MMTV vSAg-29, which is characteristic for all of the SJL RCS tumors that have been tested thus far. These results suggest that the C57L B lymphoma cells also express a similar Vβ16-specific vSAg. However, the responsiveness of a more comprehensive panel of T cell hybrids must be tested to examine the possibility that other vSAgs present in C57L mice (*Mtv8*,9,11,17,29)[131] may also be expressed by the C57L B lymphoma cells. Examination of Northern blots prepared with RNA from NJ123 cells and probed with MMTV-LTR showed strong expression of the 1.8 mRNA for MMTV-LTR, which is the size of the mRNA found in SJL lymphomas to encode the vSAg-29, but the identity of the MMTV-LTR expressed by the C57L lymphomas has not yet been further established.

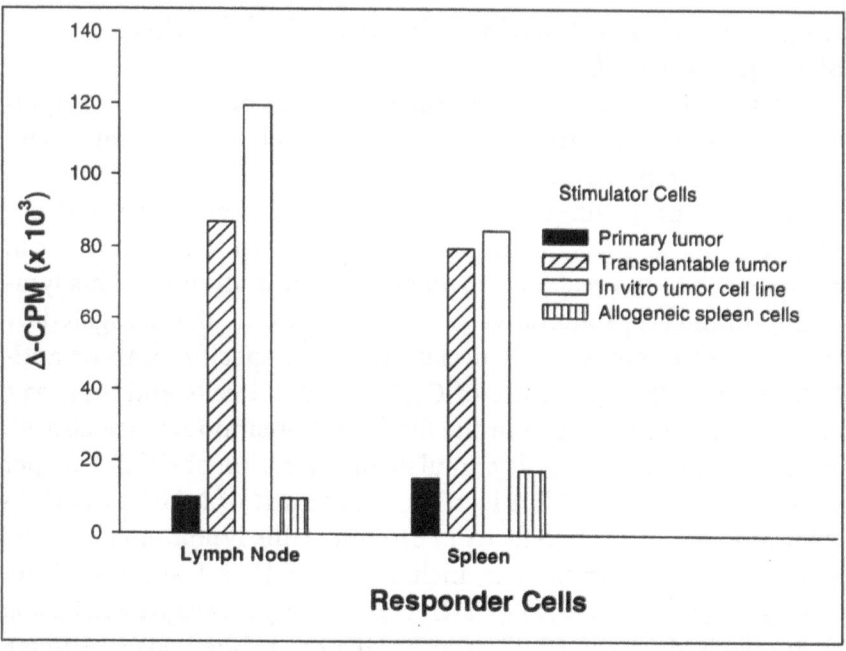

Fig. 5.2. Lymph node or spleen responder cells were taken from young (approx. 8 week-old) C57L/J donors and were cultured for 96 hours with irradiated stimulator cells. Cultures were pulsed with ³H-thymidine for the last 18 hours of culture and then harvested onto glass fiber filters and counted in a scintillation counter. NJ123 is a primary tumor that arose in the mesenteric lymph node of an aging (approx. 2 year-old) C57L/J mouse. Transplantable NJ123 cells were taken from the tumorous mesenteric lymph node of a C57L/J mouse injected approximately 3 months earlier with NJ123 primary tumor cells. cNJ123.5 is an in vitro cell line derived from NJ123.

In Vivo Growth Requirements of C57L Lymphomas

Although the C57L B cell lymphomas possess many properties in common with the SJL lymphomas, major differences in the frequency and age of onset exist. While the SJL lymphomas arise in about 90% of mice by 13 months of age, less than 50% of C57L mice develop lymphomas, and these arise much later in life (>20 months of age). One difference between C57L and SJL mice that may contribute to the lower frequency and later onset of lymphomas in C57L mice is the high natural killer (NK) cell phenotype of C57L[132,133] compared to the low NK activity in SJL/J mice.[134]

To investigate the possibility that NK cells influence lymphoma growth, young C57L recipients were injected with a polyclonal anti-NK rabbit antiserum (anti-AsialoGM₁) 24 hours before transfer of NJ123 lymphoma cells. As presented in Table 5.5, mice that received anti-NK antibody showed significant tumor growth in spleen (ex-

Table 5.4. IL-2 production by Vβ16⁺ T-T hybrids in response to C57L and SJL lymphoma cells

Stimulator Cells (Strain)	Ratio Responder/ Stimulator Cells [a]	IL-2 Production[a] (pg/ml) by	
		1D1-E7 (Vβ16)[b]	SK23-7.4 (Vβ17a)[b]
γ-cNJ123.5 (C57L)	4:1	1,672	0
γ-cNJ123.5 (C57L)	10:1	437	0
γ-cNJ117 (SJL)	100:1	3,388	0

a) γ-irradiated (cNJ123.5: 5,000R; NJ117: 15,000 R) in vitro lymphoma cell lines, cNJ123.5 (C57L/J-derived) or NJ117 (SJL/J-derived), were cultured for 24 hours at the indicated responder/stimulator ratio (R/S) with T-T hybridoma cells (10^5 /well) in RPMI-1640 + 10% FCS + 0.05 mM 2-mercaptoethanol. Supernatants (SN) collected after 24 hours of culture were assayed for IL-2 content by ELISA (mIL-2 Mini Kit, Endogen, Inc., Woburn, MA). All background IL-2 values for media alone controls were <3 pg/ml.

b) 1D1-E7 hybridoma cells are RCS-specific; SK23-7.4 are KLH-specific, I-Aˢ restricted.

pressed as a percentage of body weight) about 2 weeks after tumor cell injection, compared to mice given normal rabbit serum or to mice injected with tumor cells alone. Indeed, without anti-NK antibody treatment, NJ123 lymphoma cells either never grew, or else took several months to show palpable growth, in C57L recipients (data not shown).

One hallmark of the SJL lymphomas is their inability to develop or grow progressively in immunocompromised recipients because of their dependence on responsive host T helper cells for the provision of necessary tumor growth-promoting cytokines.[113,115] Since the C57L B lymphomas possess the ability to stimulate syngeneic T helper cells, it appeared that the growth of C57L lymphoma cells might also be dependent on the presence of an intact T helper cell compartment in recipient mice. The rapid growth of C57L lymphoma cells in NK-depleted recipients afforded an opportunity to determine if treatments which compromise the lymphoid system of recipient mice also inhibit their ability to support growth of the transplantable lymphoma. Mice injected with anti-NK antibody 24 hours before transfer of NJ123 lymphoma cells showed splenic tumor growth at 14 days that was >10% of their body weight. In contrast, mice that were given anti-NK antibody plus a sublethal dose of γ-irradiation (650R) 24 hours prior to injection of tumor cells had

Table 5.5. Influence of NK cells on growth of syngeneic B lymphoma in C57L/J mice

Antibody[a] Injected	Treatment	Spleen Weight[b] (gm) ± SD	Spleen as % of[b] BW ± SD	p vs. Control
None or NRS, Day 0	Tumor cells i.v. on day 1	0.24 ± 0.20	1.07 ± 0.88[c]	NS
Anti-Asialo GM₁, Day 0	"	1.67 ± 0.30	6.97 ± 0.92[c]	< 0.0001
None	No tumor cells injected	0.09 ± 0.02	0.45 ± 0.08	Control

a) Mice in all groups were injected on day 0 with transplantable tumor cells derived from a C57L-derived lymphoma, NJ123. Anti-NK Ab-treated mice were injected i.v. with anti-Asialo GM₁ antibody (100 µl) 24 hours before injection of tumor cells. Mice in the normal rabbit serum (NRS) group were injected i.v. with 100 µl of NRS 24 hours before injection of tumor cells.
b) Mice were killed between 11 and 17 days after injection of tumor cells. Values shown in the table represent the means ± SD of data from three independent experiments.
c) Significance of the difference between these two values: p < 0.0001.

splenic tumor growth that was only 25% of that seen in nonirradiated recipients over the same interval. An even greater inhibition (93%) of tumor growth was observed when anti-CD4 mAb treatment was given, in lieu of γ-irradiation, to NK-depleted, NJ123 tumor cell-injected C57L recipients (Erianne, GS and Ponzio, NM, unpublished observations).

These results clearly show that the C57L B cell lymphomas, l e their SJL counterparts are totally dependent on a host T helper cell response in order to grow progressively in vivo. Presumably, as is the case in SJL mice, this T helper cell dependence is also a requirement for the development of primary lymphomas in C57L mice. However, in C57L mice, the final appearance of malignancy is also influenced by the activity of host NK cells. Thus, the concept of "reverse immunological surveillance" is not unique to the SJL lymphomas for which the phrase was originally coined, but may be a more widespread contributing factor in the etiology of germinal center derived B cell lymphoma development, at least in the mouse.

Conclusions

The study of murine germinal center derived lymphomas suggests a close interaction between host lymphoid and lymphoma cells, resulting in negative effects from host NK cells and positive effects from host CD4 T cells on lymphoma growth. Although no such information is available for human germinal center-derived and other B cell lymphomas, infiltration by host T cells is frequently present and interaction of the lymphoma cells with those T cells is suggested, as in the case of maltomas and nodular lymphocyte predominant Hodgkin's disease. Such interaction could be the result of presentation of conventional antigen by host or tumor cells, tumor-specific antigens resulting from chromosomal translocations or mutations, and/or superantigens due to retroviral or viral protein expression by the tumor cells.

Some other characteristics, such as the expression of surface CD30, that appear to be shared by certain human and murine germinal center derived lymphomas, require further attention. The significance of the production of a variety of cytokines by both human and murine germinal center derived lymphomas also needs to be elucidated. Of particular interest in this respect, in view of their great importance for germinal center formation, are LTα, LTβ and TNF-α.

Acknowledgments

We are grateful for the expert advice from Dr. G. Frizzera (NYUMC, New York, NY) in the writing of this manuscript. We also acknowledge the permission from Dr. E.R. Podack (Univ. Miami Sch. Med., Miami, FL), Dr. J.L. Browning (Biogen, Cambridge, MA), and Dr. Z. Ovary, D. Charyton, W.J. Simmons and C. Huang (NYUMC, New York, NY) to include some of their unpublished data.

References

1. Brill NE, Baehr G, Rosenthal N. Generalized giant lymph follicle hyperplasia of lymph nodes and spleen. A hitherto undescribed type. J Amer Med Assoc 1925; 84:668-671.
2. Symmers D. Follicular lymphadenopathy with splenomegaly. A newly recognized disease of the lymphatic system. Arch Path 1927; 3:816.
3. Rappaport H, Winter WJ, Hicks EB. Follicular lymphoma. A reevaluation of its position in the scheme of malignant lymphoma, based on a survey of 253 cases. Cancer 1954; 9:792.
4. Lennert K. Germinal centers and germinal center neoplasms. Nippon Ketsueki Gakkai Zasshi 1969; 32:495-500.
5. Lennert K, Mohri N, Stein H et al. The histopathology of malignant lymphoma. Br J Haematol 1975; 31(S):193.

6. Lukes RJ, Collins RD. Immunologic characterization of human malignant lymphomas. Cancer 1974; 34(suppl):1488-1503.
7. Gallagher CJ, Gregory VM, Jones AE et al. Follicular lymphoma: Prognostic factors for response and survival. J Clin Oncol 1986; 4:1470-1480.
8. Inghirami G, Szabolcs MJ, Yee HT et al. Detection of immunoglobulin gene rearrangement of B cell non-Hodgkin's lymphomas and leukemias in fresh, unfixed and formalin-fixed, paraffin-embedded tissue by polymerase chain reaction. Lab Invest 1993; 68:746-757.
9. Stein H, Bonk A, Tolksdorf G et al. Immunohistologic analysis of the organization of normal lymphoid tissue and non-Hodgkin's lymphomas. J Histochem Cytochem 1980; 28:746-760.
10. Harris NL, Nadler LM, Bhan AK. Immunohistologic characterization of two malignant lymphomas of germinal center type (centroblastic/centrocytic and centrocytic) with monoclonal antibodies. Follicular and diffuse lymphomas of small-cleaved-cell type are related but distinct entities. Am J Pathol 1984; 117:262-272.
11. Stein H, Lennert K, Feller A et al. Immunological analysis of tissue sections in diagnosis of lymphoma, Churchill Livingstone, New York, 1985.
12. Hollema H, Poppema S. Immunophenotypes of malignant lymphoma centroblastic-centrocytic and malignant lymphoma centrocytic: An immunohistologic study indicating a derivation from different stages of B cell differentiation. Hum Pathol 1988; 19:1053-1059.
13. Stein H, Lennert K, Feller AC et al. Immunohistological analysis of human lymphoma: Correlation of histological categories. Adv Cancer Res 1984; 42:67-147.
14. Dvoretsky P, Wood GS, Levy R et al. T-lymphocyte subsets in follicular lymphomas compared with those in non-neoplastic lymph nodes and tonsils. Hum Pathol 1982; 13:618-625.
15. Harris NL, Bhan AK. Distribution of T-cell subsets in follicular and diffuse lymphomas of B-cell type. Am J Pathol 1983; 113:172-180.
16. Tsujimoto Y, Finger LR, Yunis J et al. Cloning of the chromosome breakpoint of neoplastic B cells with the t(14;18) chromosome translocation. Science 1984; 226:1097-1099.
17. Cleary ML, Smith SD, Sklar J. Cloning and structural analysis of cDNAs for bcl-2 and a hybrid bcl-2/immunoglobulin transcript resulting from the t(14:18) translocation. Cell 1986; 47:19-28.
18. Hockenbery DM, Oltvai ZN, Yin XM et al. Bcl-2 functions in an antioxidant pathway to prevent apoptosis. Cell 1993; 75:241-251.
19. Tsujimoto Y, Cossman J, Jaffe E et al. Involvement of the bcl-2 gene in human follicular lymphoma. Science 1985; 228:1440-1443.
20. Zelenetz AD, Chu G, Galili N et al. Enhanced detection of the t(14;18) translocation in malignant lymphoma using pulsed-field gel electrophoresis. Blood 1991; 78:1552-1560.
21. Nunez G, Seto M, Seremetis S et al. Growth- and tumor-promoting effects of deregulated BCL2 in human B-lymphoblastoid cells. Proc Natl Acad Sci USA 1989; 86:4589-4593.

22. Nunez G, London L, Hockenbery D et al. Deregulated Bcl-2 gene expression selectively prolongs survival of growth factor-deprived hemopoietic cell lines. J Immunol 1990; 144:3602-3610.
23. Hockenbery DM, Zutter M, Hickey W et al. BCL2 protein is topographically restricted in tissues characterized by apoptotic cell death. Proc Natl Acad Sci USA 1991; 88:6961-6965.
24. Inghirami G, Frizzera G. Role of the bcl-2 oncogene in Hodgkin's disease. Am J Clin Pathol 1994; 101:681-683.
25. Gaidano G, Ballerini P, Gong JZ et al. P53 mutations in human lymphoid malignancies, association with Burkitt lymphoma and chronic lymphocytic leukemia. Proc Natl Acad Sci 1991; 88:5413-5417.
26. Lo Coco F, Gaidano G, Louie DC et al. p53 mutations are associated with histologic transformation of follicular lymphoma. Blood 1993; 82:2289-2295.
27. Sander CA, Yano T, Clark HM et al. p53 mutation is associated with progression in follicular lymphomas. Blood 1993; 82:1994-2004.
28. Matolcsy A, Inghirami G, Knowles DM. Molecular genetic demonstration of the diverse evolution of Richter's syndrome (chronic lymphocytic leukemia and subsequent large cell lymphoma). Blood 1994; 83:1363-1372.
29. Lee JT, Innes DJ Jr, Williams ME. Sequential bcl-2 and c-myc oncogene rearrangements associated with the clinical transformation of non-Hodgkin's lymphoma. J Clin Invest 1989; 84:1454-1459.
30. Lukes RJ, Collins RD. New approaches to the classification of the lymphomata. Br J Cancer 1975; 31 SUPPL 2:1-28.
31. Ling NR, Hardie D, Lowe J et al. A phenotypic study of cells from Burkitt lymphoma and EBV-B-lymphoblastoid lines and their relationship to cells in normal lymphoid tissues. Int J Cancer 1989; 43:112-118.
32. Favrot MC, Philip I, Philip T et al. Distinct reactivity of Burkitt's lymphoma cell lines with eight monoclonal antibodies correlated with the ethnic origin. JNCI 1984; 73:841-847.
33. MacLennan IC. Endemic Burkitt's lymphomas (myc), sporadic Burkitt's lymphomas. In: AIDS Lymphomas. Vol. 103. Melchers M, ed. Basel: Roche, 1987.
34. Pelicci PG, Knowles DMd, Magrath I et al. Chromosomal breakpoints and structural alterations of the c-myc locus differ in endemic and sporadic forms of Burkitt lymphoma. Proc Natl Acad Sci USA 1986; 83:2984-2988.
35. Haluska FG, Tsujimoto Y, Croce CM. The t(8;14) chromosome translocation of the Burkitt lymphoma cell line Daudi occurred during immunoglobulin gene rearrangement and involved the heavy chain diversity region. Proc Natl Acad Sci USA 1987; 84:6835-6839.
36. Neri A, Barriga F, Knowles DM et al. Different regions of the immunoglobulin heavy-chain locus are involved in chromosomal translocations in distinct pathogenetic forms of Burkitt lymphoma. Proc Natl Acad Sci USA 1988; 85:2748-2752.

37. Shiramizu B, Barriga F, Neequaye J et al. Patterns of chromosomal breakpoint locations in Burkitt's lymphoma: Relevance to geography and Epstein-Barr virus association. Blood 1991; 77:1516-1526.

38. Ladanyi M, Offit K, Parsa NZ et al. Follicular lymphoma with t(8;14)(q24;q32): A distinct clinical and molecular subset of t(8;14)-bearing lymphomas. Blood 1992; 79:2124-2130.

39. Gibbons DL, Rowe M, Cope AP et al. Lymphotoxin acts as an autocrine growth factor for Epstein-Barr virus-transformed B cells and differentiated Burkitt lymphoma cell lines. Eur J Immunol 1994; 24:1879-1885.

40. Kiwanuka J, Marti G, Moore J et al. Immunoglobulin delta expression in Burkitt's lymphoma cell lines. J Natl Cancer Inst 1989; 81:1075-1079.

41. Mann RB, Jaffe ES, Braylan RC et al. Non-endemic Burkitts's lymphoma. A B-cell tumor related to germinal centers. N Engl J Med 1976; 295:685-691.

42. Schuurman HJ, van Baarlen J, Huppes W et al. Immunophenotyping of non-Hodkin's lymphoma. Lack of correlation between immunophenotype and cell morphology. Am J Pathol 1987; 129:140-151.

43. Mangeney M, Lingwood CA, Taga S et al. Apoptosis induced in Burkitt's lymphoma cells via Gb3/CD77, a glycolipid antigen. Cancer Res 1993; 53:5314-5319.

44. Raphael MM, Audouin J, Lamine M et al. Immunophenotypic and genotypic analysis of acquired immunodeficiency syndrome-related non-Hodgkin's lymphomas. Correlation with histologic features in 36 cases. French Study Group of Pathology for HIV-Associated Tumors. Am J Clin Pathol 1994; 101:773-782.

45. Tamaru J, Hummel M, Marafioti T et al. Burkitt's lymphomas express VH genes with a moderate number of antigen-selected somatic mutations. Am J Pathol 1995; 147:1398-1407.

46. Billaud M, Rousset F, Calender A et al. Low expression of lymphocyte function-associated antigen (LFA)-1 and LFA-3 adhesion molecules is a common trait in Burkitt's lymphoma associated with and not associated with Epstein-Barr Virus. Blood 1990; 75:1827-1833.

47. Vacca A, Ranieri G, Ribatti D et al. Differential expression of two ICAM-1 epitopes and LFA-1 chains in B-cell non-Hodgkin's lymphomas. Eur J Haematol 1994; 53:85-92.

48. Kelly DR, Nathwani BN, Griffith RC et al. A morphologic study of childhood lymphoma of the undifferentiated type. The Pediatric Oncology Group experience. Cancer 1987; 59:1132-1137.

49. Berstein JI, Coleman CN, Strickler JG et al. Combined modality therapy for adults with small noncleaved cell lymphoma (Burkitt's and non-Burkitt's types). J Clin Oncol 1986; 4:847-858.

50. Miliauskas JR, Berard CW, Young RC et al. Undifferentiated non-Hodgkin's lymphomas (Burkitt's and non-Burkitt's types). The relevance of making this histologic distinction. Cancer 1982; 50:2115-2121.

51. Lardelli P, Bookman MA, Sundeen J et al. Lymphocytic lymphoma of intermediate differentiation. Morphologic and immunophenotypic spectrum and clinical correlations. Am J Surg Pathol 1990; 14:752-763.

52. Weisenburger DD, Sanger WG, Armitage JO et al. Intermediate lymphocytic lymphoma: Immunophenotypic and cytogenetic findings. Blood 1987; 69:1617-1621.

53. Frizzera G, Sakurai M, Notohara K et al. t(11;14)(q13;q32) in B-cell lymphomas (intermediately differentiated lymphocytic and follicular). A report of four cases. Am J Clin Pathol 1991; 95:684-691.

54. Vandenberghe E, De Wolf-Peeters C, Wlodarska I et al. Chromosome 11q rearrangements in B non Hodgkin's lymphoma. Br J Haematol 1992; 81:212-217.

55. Nizze H, Cogliatti SB, von Schilling C et al. Monocytoid B-cell lymphoma: Morphological variants and relationship to low grade B-cell lymphoma of the mucosa-associated lymphoid tissue. Histopathology 1991; 18:403-414.

56. MacLennan IC, Gray D. Antigen-driven selection of virgin and memory B cells. Immunol Rev 1986; 91:61-85.

57. Dunn-Walters DK, Isaacson PG, Spencer J. Analysis of mutations in immunoglobulin heavy chain variable region genes of microdissected marginal zone (MGZ) B cells suggests that the MGZ of human spleen is a reservoir of memory B cells. J Exp Med 1995; 182:559-566.

58. Isaacson PG. Is gastric lymphoma an infection disease? Human Pathol 1993; 24:569-570.

59. Ye BH, Chaganti S, Niu H et al. Chromosomal translocations cause deregulatiod BCL6 expression by promotor substitution in B cell lymphoma. EMBO J 1995; 14:6209-6217.

60. Gruss HJ, Brach MA, Drexler HG et al. Interleukin 9 is expressed by primary and cultured Hodgkin and Reed-Sternberg cells. Cancer Res 1992; 52:1026-1031.

61. Hu EH, Ellison D, Zovich D et al. Molecular analysis of Hodgkin's disease with abundant Reed-Sternberg cells. Hematol Pathol 1990; 4:27-35.

62. Kuppers R, Rajewsky K, Zhao M et al. Hodgkin disease: Hodgkin and Reed-Sternberg cells picked from histological sections show clonal immunoglobulin gene rearrangements and appear to be derived from B cells at various stages of development. Proc Natl Acad Sci USA 1994; 91:10962-10966.

63. Roth MS, Schnitzer B, Bingham EL et al. Rearrangement of immunoglobulin and T-cell receptor genes in Hodgkin's disease. Am J Pathol 1988; 131:331-338.

64. Marafioti T, Hummel M, Anagnostopoulos I et al. Origin of nodular lymphocyte-predominant Hodgkin's disease from a clonal expansion of highly mutated germinal-center B cells. N Engl J Med 1997; 337:453-458.

65. Ohno T, Stribley JA, Wu G et al. Clonality in nodular lymphocyte-predominant Hodgkin's disease. N Engl J Med 1997; 337:459-465.

66. Hansmann ML, Godde-Salz E, Hui PK et al. Cytogenetic findings in nodular paragranuloma (Hodgkin's disease with lymphocytic predominance; nodular) and in progressively transformed germinal centers. Cancer Genet Cytogenet 1986; 21:319-325.

67. Stetler-Stevenson M, Crush-Stanton S, Cossman J. Involvement of the bcl-2 gene in Hodgkin's disease. J Natl Cancer Inst 1990; 82:855-858.

68. Thangavelu M, Le Beau MM. Chromosomal abnormalities in Hodgkin's disease. Hematol Oncol Clin North Am 1989; 3:221-236.

69. Durkop H, Anagnostopoulos I, Bulfone-Paus S et al. Expression of several members of the TNF-ligand and receptor family on tonsillar B cells. Br J Haematol 1997; 98:863-868.

70. Amakawa R, Hakem A, Kundig TM et al. Impaired negative selection of T cells in Hodgkin's disease antigen CD30-deficient mice. Cell 1996; 84:551-562.

71. Gattei V, Degan M, Gloghini A et al. CD30 ligand is frequently expressed in human hematopoietic malignancies of myeloid and lymphoid origin. Blood 1997; 89:2048-2059.

72. Kuhn-Hallek I, Sage DR, Stein L et al. Expression of recombination activating genes (RAG-1 and RAG-2) in Epstein-Barr virus-bearing B cells. Blood 1995; 85:1289-1299.

73. Knecht H, Brousset P, Bachmann E et al. Expression of human recombination activating genes (RAG-1 and RAG-2) in lymphoma. Leuk Lymphoma 1994; 15:399-403.

74. Knecht H. Recombination activating gene-expression in lymphoma cell lines and lymphomas. Blood 1995; 86:412-414.

75. Hikida M, Mori M, Takai T et al. Reexpression of RAG-1 and RAG-2 genes in activated mature mouse B cells. Science 1996; 274:2092-2094.

76. Han S, Zheng B, Schatz DG et al. Neoteny in lymphocytes: Rag1 and Rag2 expression in germinal center B cells. Science 1996; 274: 2094-2097.

77. Papavasiliou F, Casellas R, Suh H et al. V(D)J recombination in mature B Cells: A mechanism for altering antibody responses. Science 1997; 278:298-301.

78. Ye BH, Lista F, Lo Coco F et al. Alterations of a zinc finger-encoding gene, Bcl-6, in diffuse large-cell lymphoma. Science 1993; 262:747-750.

79. Lo Coco F, Ye BH, Lista F et al. Rearrangements of the BCL6 gene in diffuse large cell non-Hodgkin's lymphoma. Blood 1994; 83: 1757-1759.

80. Bastard C, Deweindt C, Kerkaert JP et al. LAZ3 rearrangements in non-Hodgkin's lymphoma. Correlation with histology, immunophenotype, karyotype, clinical outcome in 217 patients. Blood 1994; 83:2423-2427.

81. Cattoretti G, Chang CC, Cechova K et al. Bcl-6 protein is expressed in germinal-center B cells. Blood 1995; 86:45-53.

82. Onizuka T, Moriyama M, Yamochi T et al. Bcl-6 gene product, a 92- to 98-kD nuclear phosphoprotein, is highly expressed in germinal center B cells and their neoplastic counterparts. Blood 1995; 86:28-37.

83. Umetsu DT, Esserman L, Donlon TA et al. Induction of proliferation of human follicular (B type) lymphoma cells by cognate interaction with CD4$^+$ T cell clones. J Immunol 1990; 144:2550-2557.
84. Macon WR, Williams ME, Greer JP et al. T-cell-rich B-cell lymphomas. A clinicopathologic study of 19 cases. Am J Surg Pathol 1992; 16:351-363.
85. Baddoura FK, Chan WC, Masih AS et al. T-cell-rich B-cell lymphoma. A clinicopathologic study of eight cases. Am J Clin Pathol 1995; 103:65-75.
86. Dolcetti R, De Re V, Carbone A et al. Genotypic and immunohistological demonstration of the progression of an unusual reactive-like B-cell lymphoproliferative disorder to a high grade diffuse lymphoma. Human Pathol 1995; 26:348-354.
87. Poppema S. The nature of the lymphocytes surrounding Reed-Sternberg cells in nodular lymphocyte predominance and in other types of Hodgkin's disease. Am J Pathol 1989; 135:351-357.
88. Kamel OW, Gelb AB, Shibuya RB et al. Leu 7 (CD57) reactivity distinguishes nodular lymphocyte predominance Hodgkin's disease from nodular sclerosing Hodgkin's disease, T-cell-rich B-cell lymphoma and follicular lymphoma. Am J Pathol 1993; 142:541-546.
89. Poppema S. Immunology of Hodgkin's disease. Baillieres Clin Haematol 1996; 9:447-457.
90. Andersson E, Ohlin M, Borrebaeck CA et al. CD4$^+$CD57$^+$ T cells derived from peripheral blood do not support immunoglobulin production by B cells. Cell Immunol 1995; 163:245-253.
91. Falini B, Bigerna B, Pasqualucci L et al. Distinctive expression pattern of the Bcl-6 protein in nodular lymphocyte predominance Hodgkin's disease. Blood 1996; 87:465-471.
92. Gruss HJ, Herrmann F, Gattei V et al. CD40/CD40 ligand interactions in normal, reactive and malignant lympho-hematopoietic tissues. Leuk Lymphoma 1997; 24:393-422.
93. Wotherspoon AC. Gastric MALT lymphoma and Helicobacter pylori. Yale J Biol Med 1997; 69:61-68.
93a. Hussell T, Isaacson PG, Crabtree JE et al. The response of cells from low-grade B-cell gastric lymphomas of mucosa-associated lymphoid tissue to Helicobacter pylori. Lancet 1993; 342:571-574.
93b. Hussell T, Isaacson PG, Crabtree JE et al. Helicobacter pylori-specific tumour-infiltrating T cells provide contact dependent help for the growth of malignant B cells in low-grade gastric lymphoma of mucosa-associated lymphoid tissue. J Pathol 1996; 178:122-127.
94. Singh N, Wright DH. The value of immunohistochemistry on paraffin wax embedded tissue sections in the differentiation of small lymphocytic and mantle cell lymphomas. J Clin Pathol 1997; 50:16-21.
95. Hirokawa M, Lee M, Kitabayashi A et al. Autoreactive T cell-dependent polyclonal hypergammaglobulinemia in mantle cell lymphoma. Leuk Lymphoma 1994; 14:509-513.

96. Serrano D, Monteiro J, Allen SL et al. Clonal expansion within the CD4⁺CD57⁺ and CD8⁺CD57⁺ T cell subsets in chronic lymphocytic leukemia. J Immunol 1997; 158:1482-1489.
97. Osterborg A, Werner A, Halapi E et al. Clonal CD8⁺ and CD52⁻ T cells are induced in responding B cell lymphoma patients treated with Campath-1H (anti-CD52). Eur J Haematol 1997; 58:5-13.
98. Siegler R, Rich MA. Pathogenesis of reticulum cell sarcoma in mice. J Natl Cancer Inst 1968; 41:125-143.
99. Pattengale PK, Taylor CR. Experimental models of lymphoproliferative disease. The mouse as a model for human non-Hodgkin's lymphomas and related leukemias. Am J Pathol 1983; 113:237-265.
100. Ponzio NM, Tsiagbe VK, Thorbecke GJ. Superantigens related to B cell hyperplasia. Springer Semin Immunopathol 1996; 17:285-306.
101. Secord EA, Edington JM, Thorbecke GJ. The Emu-bcl-2 transgene enhances antigen-induced germinal center formation in both BALB/c and SJL mice but causes age-dependent germinal center hyperplasia only in the lymphoma-prone SJL strain. Am J Pathol 1995; 147:422-433.
102. Seibert K, Pollard M, Nordin A. Some aspects of humoral immunity in germ-free and conventional SJL-J mice in relation to age and pathology. Cancer Res 1974; 34:1707-1719.
103. Stavnezer J, Lasky JL, Ponzio NM et al. Reticulum cell sarcomas of SJL mice have rearranged immunoglobulin heavy and light chain genes. Eur J Immunol 1989; 19:1063-1069.
104. Tsiagbe VK, Thorbecke GJ. Paraproteins and primary lymphoma in SJL mice. I. Individuality of idiotypes on paraproteins. Cell Immunol 1990; 129:494-502.
105. Tsiagbe VK, Asakawa J, Ponzio NM et al. Paraproteins and primary lymphoma in SJL mice. II. Primary lymphomas do not produce paraprotein. Cell Immunol 1990; 129:503-512.
106. Carswell EA, Lerman SP, Thorbecke GJ. Properties of reticulum cell sarcomas in SJL/J mice. II. Fate of labeled tumor cells in normal and irradiated syngeneic mice. Cell Immunol 1976; 23:39-52.
107. Lerman SP, Carswell EA, Chapman J et al. Properties of reticulum cell sarcomas in SJL/J mice. III. Promotion of tumor growth in irradiated mice by normal lymphoid cells. Cell Immunol 1976; 23:53-67.
108. Katz IR, Chapman-Alexander J, Jacobson EB et al. Growth of SJL/J derived transplantable reticulum cell sarcoma as related to its ability to induce T-cell proliferation in the host. III. Studies on thymectomized and congenitally athymic SJL mice. Cell Immunol 1981; 65:84-92.
109. Katz IR, Chapman-Alexander J, Thorbecke GJ. Growth of SJL/J-derived transplantable reticulum cell sarcoma as related to its ability to induce T-cell proliferation in the host. IV. Effect of thymectomy on primary lymphoma incidence. Cell Immunol 1982; 74:394-397.
110. Alisauskas RM, Ponzio NM. T helper cell-specific monoclonal antibody inhibits progressive growth of B cell lymphomas in syngeneic SJL/J mice. Cell Immunol 1989; 119:286-303.

111. Alisauskas RM, Friedman CA, Ponzio NM. Influence of T helper cell specific monoclonal antibody on progressive growth of B cell lymphomas in SJL/J mice. Cancer Commun 1990; 2:33-43.

112. Lasky JL, Ponzio NM, Thorbecke GJ. Characterization and growth factor requirements of SJL lymphomas. I. Development of a B cell growth factor-dependent in vitro cell line, cRCS-X. J Immunol 1988; 140:679-687.

113. Ponzio NM, Brown PH, Thorbecke GJ. Host-tumor interactions in the SJL lymphoma model. Int Rev Immunol 1986; 1:273-301.

114. Lasky JL, Thorbecke GJ. Characterization and growth factor requirements of SJL lymphomas. II. Interleukin 5 dependence of the in vitro cell line, cRCS-X, and influence of other cytokines. Eur J Immunol 1989; 19:365-371.

115. Ponzio NM, Zhang DJ, Tsiagbe VK et al. Influence of MTV superantigen on B cell lymphoma development in SJL/J mice. In: Tomonori K, ed. Viral Superantigens. Chapter 17. Boca Raton, FL: CRC Press, 1997:219-231.

116. Santambrogio L, Crisi GM, Leu J et al. Tolerogenic forms of autoantigens and cytokines in the induction of resistance to experimental allergic encephalomyelitis. J Neuroimmunol 1995; 58:211-222.

117. Matsumoto M, Fu YX, Molina H et al. Lymphotoxin-alpha-deficient and TNF receptor-I-deficient mice define developmental and functional characteristics of germinal centers. Immunol Rev 1997; 156:137-144.

118. DeKruyff RH, Brown PH, Thorbecke GJ et al. Characterization of SJL T cell clones responsive to syngeneic lymphoma (RCS): RCS-specific clones are stimulated by activated B cells. J Immunol 1985; 135:3581-3586.

119. Tsiagbe VK, Asakawa J, Miranda A et al. Syngeneic response to SJL follicular center B cell lymphoma (reticular cell sarcoma) cells is primarily in V beta 16+ CD4$^+$ T cells. J Immunol 1993; 150:5519-5528.

120. Tsiagbe VK, Yoshimoto T, Asakawa J et al. Linkage of superantigen-like stimulation of syngeneic T cells in a mouse model of follicular center B cell lymphoma to transcription of endogenous mammary tumor virus. Embo J 1993; 12:2313-2320.

121. Zhang DJ, D' Eustachio P, Thorbecke GJ. The Mtv29 gene encoding endogenous lymphoma superantigen in SJL mice, mapped to proximal chromosome 6. Immunogenetics 1997; 46:163-166.

122. Wheeler DA, Butel JS, Medina D et al. Transcription of mouse mammary tumor virus: Identification of a candidate mRNA for the long terminal repeat gene product. J Virol 1983; 46:42-49.

123. van Ooyen AJ, Michalides RJ, Nusse R. Structural analysis of a 1.7-kilobase mouse mammary tumor virus-specific RNA. J Virol 1983; 46:362-370.

124. Zhang DJ, Tsiagbe VK, Huang C et al. Control of endogenous mouse mammary tumor virus superantigen expression in SJL lymphomas by a promoter within the env region. J Immunol 1996; 157:3510-3517.

125. Miller CL, Garner R, Paetkau V. An activation-dependent, T-lymphocyte-specific transcriptional activator in the mouse mammary tumor virus env gene. Mol Cell Biol 1992; 12:3262-3272.
126. Heston W. Genetics of neoplasia. In: Burdette W, Ed. Methodology in Mammalian Genetics. San Francisco: Holden-Day, 1963:247-268.
127. Dunn TB, Deringer MK. Reticulum cell neoplasm, type B, or the "Hodgkin's-like lesion" of the mouse. J Natl Cancer Inst 1968; 40:771-821.
128. Jones PP, Murphy DB, McDevitt HO. Variable synthesis and expression of E alpha and Ae (E beta) Ia polypeptide chains in mice of different H-2 haplotypes. Immunogenetics 1981; 12:321-337.
129. Behlke MA, Chou HS, Huppi K et al. Murine T-cell receptor mutants with deletions of beta-chain variable region genes. Proc Natl Acad Sci USA 1986; 83:767-771.
130. Wanebo HJ, Gallmeier WM, Boyse EA et al. Paraproteinemia and reticulum cell sarcoma in an inbred mouse strain. Science 1966; 154:901-903.
131. Lee BK, Eicher EM. Segregation patterns of endogenous mouse mammary tumor viruses in five recombinant inbred strain sets. J Virol 1990; 64:4568-4572.
132. Petranyi G, Kiessling R, Klein G. Genetic control of "natural" killer lymphocytes in the mouse. Immunogenetics 1975; 2:53-61.
133. Petranyi G, Kiessling R, Povey G et al. The genetic control of natural killer cell activity and its association with in vivo resistance against a Moloney lymphoma isograft. Immunogenetics 1976; 3:15-28.
134. Fitzgerald KL, Ponzio NM. Natural killer cell activity in reticulum cell sarcomas (RCS) of SJL/J mice. Cell Immunol 1979; 43:185-191.
135. Nguyen PL, Harris NL, Ritz J et al. Expression of CD95 antigen and Bcl-2 protein in non-Hodgkin's lymphomas and Hodgkin's disease. Am J Pathol 1996; 148:847-853.
136. Krajewski S, Gascoyne RD, Zapata JM et al. Immunolocalization of the ICE/Ced-3-family protease, CPP32 (Caspase-3), in non-Hodgkin's lymphomas, chronic lymphocytic leukemias, and reactive lymph nodes. Blood 1997; 89:3817-3825.
137. Hermann M, Scholman HJ, Marafioti T et al. Differential expression of apoptosis, Bcl-x and c-MYC in normal and malignant lymphoid tissues. Eur J Haematol 1997; 59:20-30.
138. Wang T, Lasota J, Hanau CA et al. Bcl-2 oncoprotein is widespread in lymphoid tissue and lymphomas but its differential expression in benign versus malignant follicles and monocytoid B-cell proliferations is of diagnostic value. Apmis 1995; 103:655-662.
139. Carbone A, Gaidano G, Gloghini A et al. Bcl-6 protein expression in AIDS-related non-Hodgkin's lymphomas: Inverse relationship with Epstein-Barr virus-encoded latent membrane protein-1 expression. Am J Pathol 1997; 150:155-165.
140. Flenghi L, Bigerna B, Fizzotti M et al. Monoclonal antibodies PG-B6a and PG-B6p recognize, respectively, a highly conserved and a formol-resistant epitope on the human Bcl-6 protein amino-terminal region. Am J Pathol 1996; 148:1543-1555.

141. Miura I, Ohshima A, Chubachi A et al. BCL6 rearrangement in a patient with mantle cell lymphoma. Ann Hematol 1997; 74:247-250.

142. Nakata M, Matsuno Y, Takenaka T et al. B-cell lymphoma accompanying monoclonal macroglobulinemia with features suggesting marginal zone B-cell lymphoma. Int J Hematol 1997; 65:405-411.

143. Kanzler H, Kuppers R, Hansmann ML et al. Hodgkin and Reed-Sternberg cells in Hodgkin's disease represent the outgrowth of a dominant tumor clone derived from (crippled) germinal center B cells. J Exp Med 1996; 184:1495-1505.

144. Braeuninger A, Kuppers R, Strickler JG et al. Hodgkin and Reed-Sternberg cells in lymphocyte predominant hodgkin disease represent clonal populations of germinal center-derived tumor B cells. Proc Natl Acad Sci USA 1997; 94:9337-9342.

145. Bahler DW, Miklos JA, Swerdlow SH. Ongoing Ig gene hypermutation in salivary gland mucosa-associated lymphoid tissue-type lymphomas. Blood 1997; 89:3335-3344.

146. Klein U, Kuppers R, Rajewsky K. Evidence for a large compartment of IgM-expressing memory B cells in humans. Blood 1997; 89: 1288-1298.

147. Oscier DG, Thompsett A, Zhu D et al. Differential rates of somatic hypermutation in V(H) genes among subsets of chronic lymphocytic leukemia defined by chromosomal abnormalities. Blood 1997; 89:4153-4160.

148. Vescio RA, Cao J, Hong CH et al. Myeloma Ig heavy chain V region sequences reveal prior antigenic selection and marked somatic mutation but no intraclonal diversity. J Immunol 1995; 155:2487-2497.

149. Baker BW, Deane M, Gilleece MH et al. Distinctive features of immunoglobulin heavy chain variable region gene rearrangement in multiple myeloma. Leuk Lymphoma 1994; 14:291-301.

150. Bakkus MH, Van Riet I, Van Camp B et al. Evidence that the clonogenic cell in multiple myeloma originates from a pre-switched but somatically mutated B cell. Br J Haematol 1994; 87:68-74.

151. Wagner SD, Martinelli V, Luzzatto L. Similar patterns of V kappa gene usage but different degrees of somatic mutation in hairy cell leukemia, prolymphocytic leukemia, Waldenstrom's macroglobulinemia, and myeloma. Blood 1994; 83:3647-3653.

152. Kuppers R, Rajewsky K, Hansmann ML. Diffuse large cell lymphomas are derived from mature B cells carrying V region genes with a high load of somatic mutation and evidence of selection for antibody expression. Eur J Immunol 1997; 27:1398-1405.

153. Harris NL, Jaffee ES, Stein H et al. A revised European-American classification of lymphoid neoplasms: A proposal from the international lymphoma study group. Blood 1994; 84:1361-1392.

154. Sole F, Woessner S, Florensa L et al. Frequent involvement of chromosomes 1, 3, 7 and 8 in splenic marginal zone B-cell lymphoma. Br J Haematol 1997; 98:446-449.

155. Drillenburg P, van der Voort R, Koopman G et al. Preferential expression of the mucosal homing receptor integrin alpha 4 beta 7 in gastrointestinal non-Hodgkin's lymphomas. Am J Pathol 1997; 150:919-927.

156. Hsu SM, Xie SS, Hsu PL et al. Interleukin-6, but not interleukin-4, is expressed by Reed-Sternberg cells in Hodgkin's disease with or without histologic features of Castleman's disease. Am J Pathol 1992; 141:129-138.
157. Merz H, Houssiau FA, Orscheschek K et al. Interleukin-9 expression in human malignant lymphomas: Unique association with Hodgkin's disease and large cell anaplastic lymphoma. Blood 1991; 78:1311-1317.
158. Goldfeld AE, Flemington EK, Boussiotis VA et al. Transcription of the tumor necrosis factor alpha gene is rapidly induced by anti-immunoglobulin and blocked by cyclosporin A and FK506 in human B cells. Proc Natl Acad Sci USA 1992; 89:12198-12201.
159. Fior R, Vita N, Raphael M et al. Interleukin-13 gene expression by malignant and EBV-transformed human B lymphocytes. Eur Cytokine Netw 1994; 5:593-600.
160. Emilie D, Zou W, Fior R et al. Production and roles of IL-6, IL-10, and IL-13 in B-lymphocyte malignancies and in B-lymphocyte hyperactivity of HIV infection and autoimmunity. Methods 1997; 11:133-142.
161. Emilie D, Coumbaras J, Raphael M et al. Interleukin-6 production in high-grade B lymphomas: Correlation with the presence of malignant immunoblasts in acquired immunodeficiency syndrome and in human immunodeficiency virus-seronegative patients. Blood 1992; 80:498-504.
162. Benjamin D, Sharma V, Knobloch TJ et al. B cell IL-7. Human B cell lines constitutively secrete IL-7 and express IL-7 receptors. J Immunol 1994; 152:4749-4757.
163. Benjamin D, Kofler G, Tschachler E. Human B-cell TNF-beta microheterogeneity. Lymphokine Cytokine Res 1992; 11:45-54.
164. Corcione A, Baldi L, Zupo S et al. Spontaneous production of granulocyte colony-stimulating factor in vitro by human B-lineage lymphocytes is a distinctive marker of germinal center cells. J Immunol 1994; 153:2868-2877.
165. Pistoia V, Cozzolino F, Rubartelli A et al. In vitro production of interleukin 1 by normal and malignant human B lymphocytes. J Immunol 1986; 136:1688-1692.
166. Zupo S, Perussia B, Baldi L et al. Production of granulocyte-macrophage colony-stimulating factor but not IL-3 by normal and neoplastic human B lymphocytes. J Immunol 1992; 148:1423-1430.
167. Corcione A, Corrias MV, Daniele S et al. Expression of granulocyte colony-stimulating factor and granulocyte colony-stimulating factor receptor genes in partially overlapping monoclonal B-cell populations from chronic lymphocytic leukemia patients. Blood 1996; 87:2861-2869.
168. Pistoia V. Production of cytokines by human B cells in health and disease. Immunol Today 1997; 18:343-350.

COLOR FIGURES

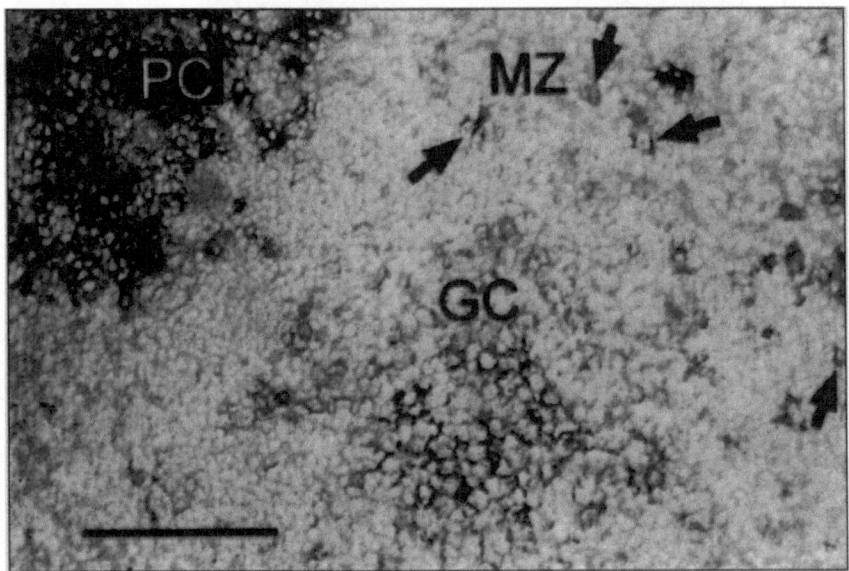

Fig. 1.2. Location of antigen-specific B cells in the spleen of a mouse in response to i.v. challenge with antigen. This B10A mouse was primed i.p. with chicken γ-globulin (CGG), given together with *B. pertussis,* and boosted with nitrophenyl (NP)-CGG. CGG-specific memory B cells are shown in blue (arrow) in the marginal zone (MZ); NP-specific cells are shown in brown. The edge of a red pulp focus of plasma cells (PC) is shown (top left). Immune complexes (black) are present on the FDCs in the GC in the center. Bar = 100 μm. Reprinted with permission from Toellner KM et al, J Exp Med 1996; 183:2303-2312.

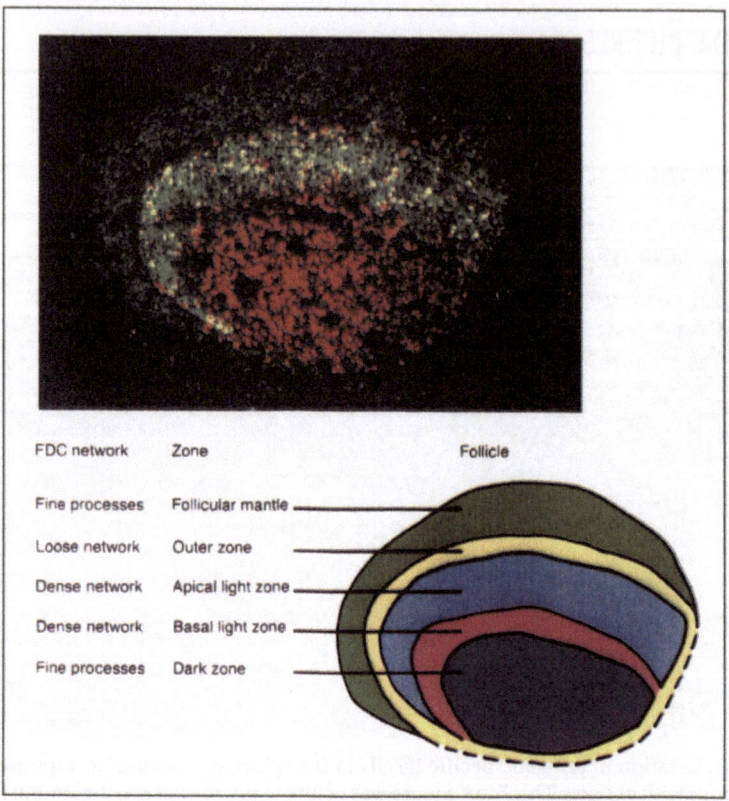

FDC network	Zone	Follicle
Fine processes	Follicular mantle	
Loose network	Outer zone	
Dense network	Apical light zone	
Dense network	Basal light zone	
Fine processes	Dark zone	

Fig. 1.3. Polarization of a germinal center in human tonsil, at a time when B cell blasts (in exponential growth) have filled the follicular dendritic cell (FDC) network. Upper part: A secondary follicle with a well-developed GC is shown in which the zonal pattern, typical of the tonsil GC, is apparent. Green immunofluorescence identifies weak CD23 expression by B cells in the follicular mantle and strong CD23 expression by FDCs in the apical light zone. Red immunofluorescence shows the heavy concentration of cells in cell cycle in the dark zone identified by the Ki67 monoclonal antibody, which identifies dividing cells. The unstained area between the follicular mantle and apical light zone is the outer zone and that between the apical light zone and the dark zone is the basal light zone. Both the outer zone and basal light zone contain occasional Ki67[+] cells. Lower part: These regions are shown with a color key. Reprinted with permission from Liu YJ et al, Immunol Today 1992; 13:18.

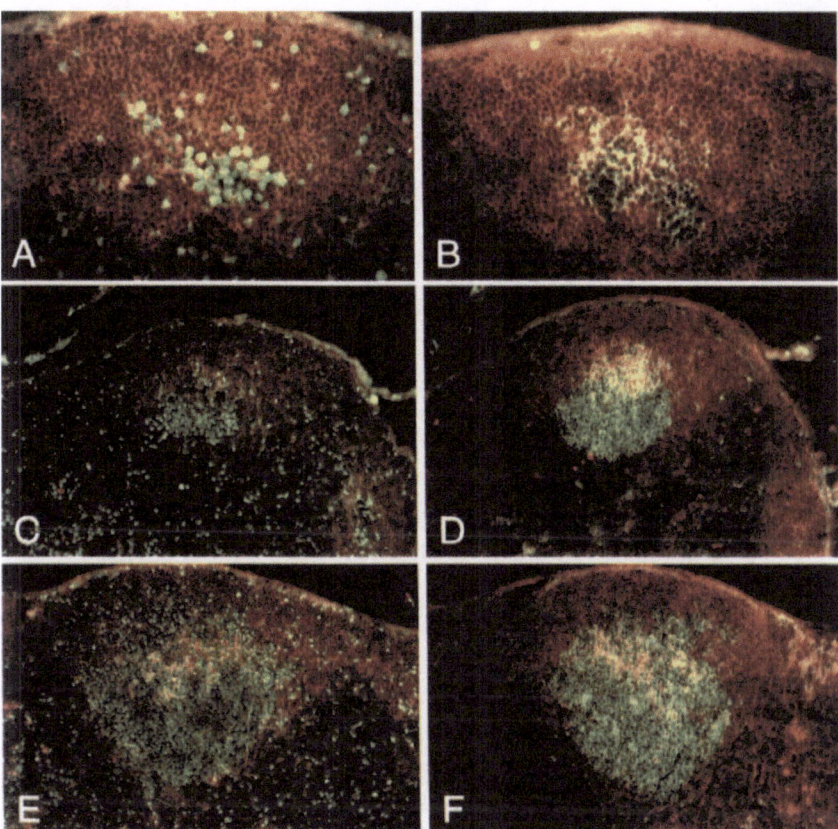

Fig. 1.4. Relationship between dividing cells, PNA$^+$ cells, and the FDC network during the GC reaction. Serial cryosections were prepared from murine lymph nodes obtained 4, 6 and 9 days post primary immunization with aluminum-precipitated ovalbumin. In all panels, IgM$^+$ cells are red. In the left panels (A,C,E), green nuclei correspond to Ki-67$^+$ dividing cells. In right panels (B,D,F), green cells correspond to peanut agglutinin (PNA)$^+$ GC cells. The FDC network in the light zone becomes visible on days 6-9, stained red due to immune complexes (ICs) recognized by the anti-IgM reagent. Note that, by day 4, Ki-67$^+$ nuclei localize as a small focus of dividing cells in the IgM$^+$ primary follicle. These cells are PNA$^+$ (panel B) and within the FDC network. By day 6, the GC size is expanding and the majority of the Ki-67$^+$ nuclei (panel C) are limited to the dark zone, while the FDC network is organized mainly in the light zone (panel D). By day 9, the GC has reached full proportion and many of the cells within the FDC network have Ki-67$^+$ nuclei (panels E & F). Magnification: day 4 A-B: x250; day 6 C-D and day 9 E-F: x90. Reprinted with permission from Kosco-Vilbois MH et al, Immunol Today 1997; 18:225-230. Photomicrographs kindly provided by Dr. M. Kosco-Vilbois.

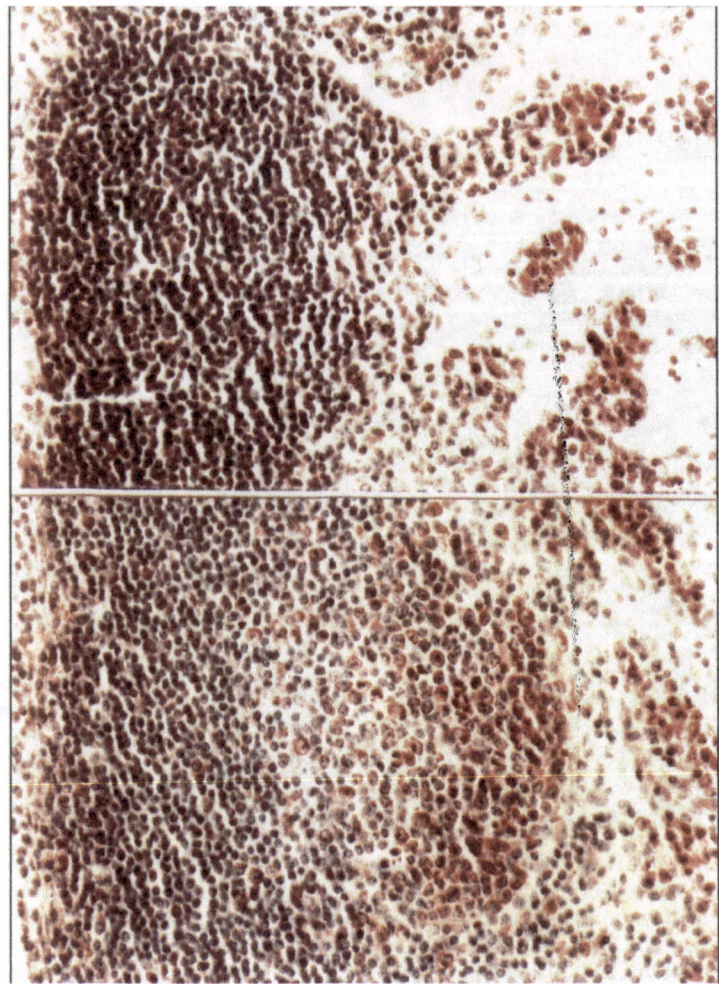

Fig. 1.6. Reconstitution of GC formation in lymph nodes from athymic mice by injection of T cells. 10^7 thymus cells were injected i.p. within the first two weeks after birth and the tissues were examined 1-2 months later. Mice were injected with antigen (killed *B. abortus* vaccine) in the front footpads 10 days before the brachial (draining) lymph nodes, shown here, were taken. Note the total absence of GCs, but presence of normal primary B cell follicles in the cortex and of plasma cells in the medullary cords in the mouse which did not receive thymus cells (top panel). In contrast, the mouse which had received thymus cells produced the typical GCs, of which one is shown here (bottom panel), that are also induced in normal mice by immunization with *B. abortus*, even though the paracortex was still relatively devoid of T cells. Illustration of observations reported on previously by Jacobson et al.[108]

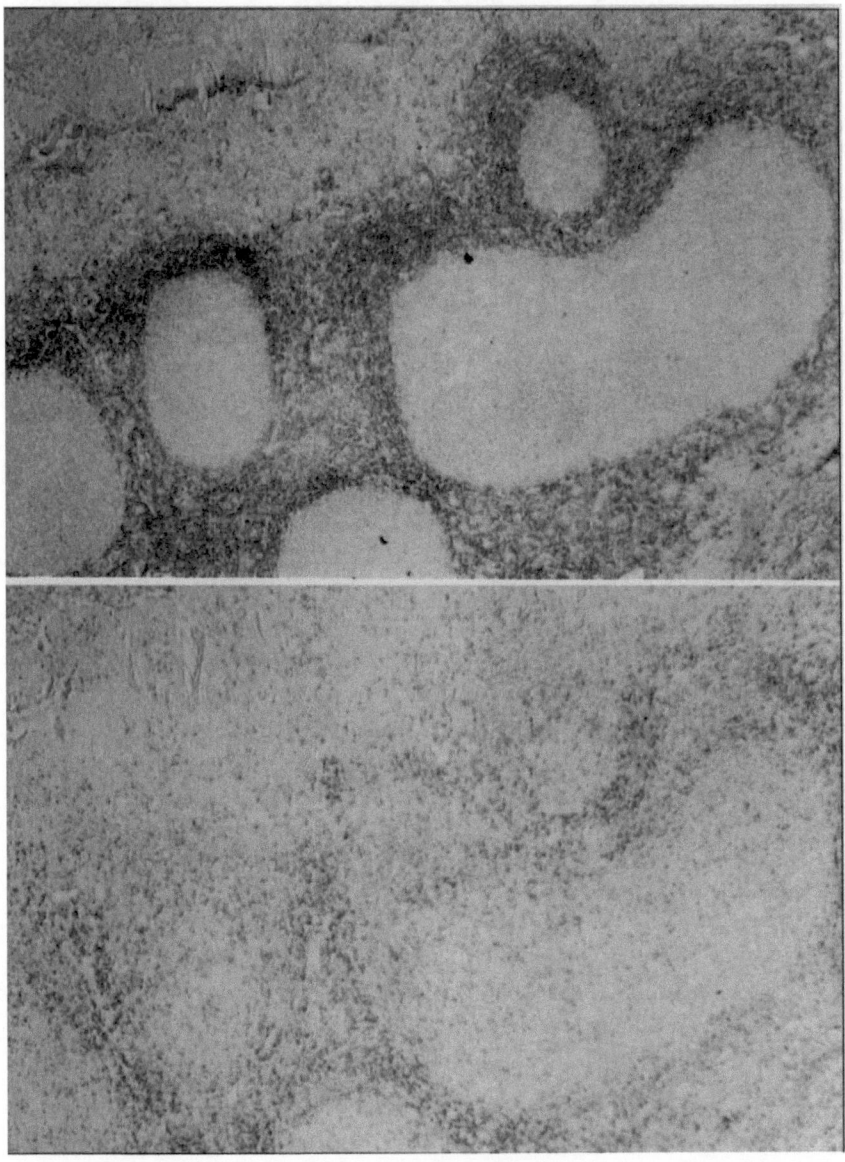

Fig. 3.1. Serial sections of a tonsil were immunohistochemically stained (brick-red) for L-selectin (top panel) or for CD3 (bottom panel) and counterstained (green) with methyl green pyronin. In the top panel, germinal centers appear as unstained areas circumscribed by L-selectin[+] cells. In the U shaped germinal center, dark and light zones appear as the solid green (bottom) or light green (top) stained areas. In the bottom panel, CD3[+] cells are in the light zone of the germinal centers. In the bottom panel, mantle zones containing the resting B cells appear as intensely green areas on top of germinal centers.

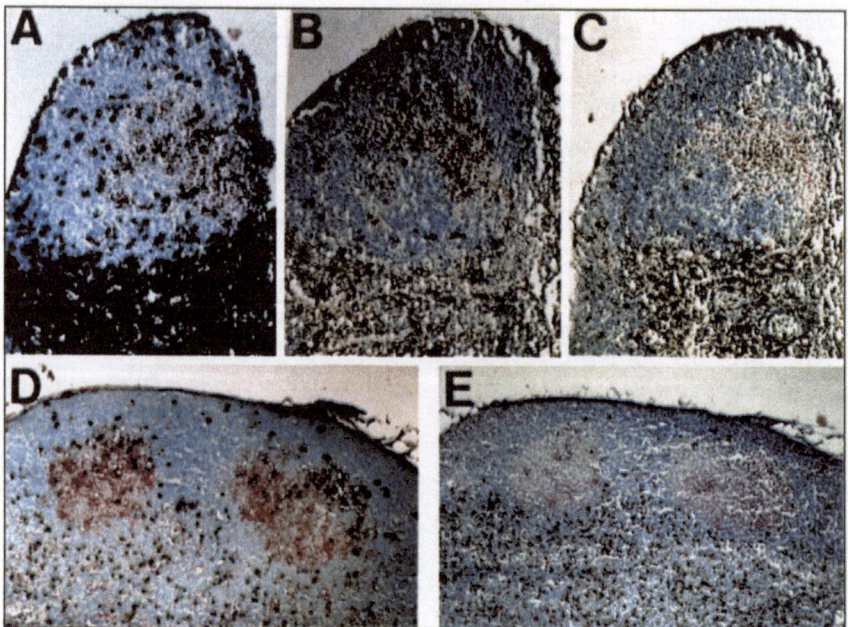

Fig. 3.2. T cells in germinal centers induced with cytochrome c. Cells in popliteal lymph node section were stained black with anti-Thy1.2 antibody (A) or with anti-Vα11 antibody (B and D), or with anti-Vβ8 antibody (C and E). Germinal centers are visualized as pink areas in the lymph node, which is counterstained green with methyl green. A, B, and C were adjacent sections from one lymph node, and D and E were from another. The popliteal lymph node was harvested 9 days (A to C) or 7 days (D and E) after immunization. The following magnifications were used. A to C x100, D and E x40. Reprinted with permission from Fuller et al, J Immunol 1993; 151:4505-4512. Copyright 1993, The American Association of Immunologists.

Index